SMART CITIES TO SMART SOCIETIES

This book explores the governance of smart cities from a holistic approach, arguing that the creation of smart cities must consider the specific circumstances of each country to improve the preservation, revitalisation, liveability, and sustainability of urban areas. The recent push for smart cities is part of an effort to reshape urban development through megaprojects, centralised master planning, and approaches that convey modernism and global affluence. However, moving towards a citywide smart transition is a major undertaking, and complexities are expected to grow exponentially. This book argues that a comprehensive approach is necessary to consider all relevant aspects. The chapters seek to identify the potential and pitfalls of the smart transformation of urban communities and its role in sustainability goals; share state-of-the-art practices concerning technology, policy, and social science dimensions in smart cities and communities; and develop opportunities for cooperation and partnership in wider and larger research and development programmes. Divided into three parts, the first part of the book highlights the significance of various societal elements and factors in facilitating a successful smart transition, with a particular emphasis on the role of human capital. The second part delves into the challenges associated with technology and its integration into smart city initiatives. The final part of the book examines the current state of regulations and policies governing smart cities. The book will be an important asset for students and researchers studying law, engineering, political science, international relations, geopolitics, economics, and engineering.

Esmat Zaidan is the Associate Dean for Academic Affairs and Associate Professor at the College of Public Policy, Hamad Bin Khalifa University, Qatar Foundation. She has published over 65 articles, predominantly in Q1 journals indexed by Scopus and ISI Web of Science, and has been recognised as one of the global 2% Most Cited Scientists in her field.

Imad Antoine Ibrahim is an Assistant Professor in Environmental Law and Governance at Department of Governance and Technology for Sustainability, Faculty of Behavioural, Management and Social Sciences, University of Twente, Enschede, the Netherlands.

Elie Azar is a Canada Research Chair and Associate Professor of Civil and Environmental Engineering at Carleton University in Ottawa, ON, Canada. His primary research focuses on integrating research methods and knowledge from different disciplines to promote a sustainable, resilient, and people-centric built environment.

Routledge Studies in Sustainability

Waste and Discards in the Asia Pacific Region
Social and Cultural Perspectives
Edited by Viktor Pál and Iris Borowy

Digital Innovations for a Circular Plastic Economy in Africa
Edited by Muyiwa Oyinlola and Oluwaseun Kolade

Critical Sustainability Sciences
Intercultural and Emancipatory Perspectives
Edited by Stephan Rist, Patrick Bottazzi and Johanna Jacobi

Transdisciplinary Research, Sustainability, and Social Transformation
Governance and Knowledge Co-Production
Tom Dedeurwaerdere

Digital Technologies for Sustainable Futures
Promises and Pitfalls
Edited by Chiara Certomà, Fabio Iapaolo, and Federico Martellozzo

Rethinking the Concept of Waste and Mass Consumption
Preserving Resources through Reuse, Repair and Recycling
Richard Waite

Mediating Sustainability in the Consumer Society
Edited by Astrid Skjerven, Lisbeth Løvbak Berg, Liv Merete Nielsen, and Dagny Stuedahl

Smart Cities to Smart Societies
Moving Beyond Technology
Edited by Esmat Zaidan, Imad Antoine Ibrahim, and Elie Azar

For more information on this series, please visit: www.routledge.com/Routledge-Studies-in-Sustainability/book-series/RSSTY

SMART CITIES TO SMART SOCIETIES

Moving Beyond Technology

Edited by Esmat Zaidan, Imad Antoine Ibrahim, and Elie Azar

Routledge
Taylor & Francis Group
LONDON AND NEW YORK

earthscan
from Routledge

Designed cover image: © Shutterstock

First published 2025
by Routledge
4 Park Square, Milton Park, Abingdon, Oxon OX14 4RN

and by Routledge
605 Third Avenue, New York, NY 10158

Routledge is an imprint of the Taylor & Francis Group, an informa business

British Library Cataloguing-in-Publication Data
A catalogue record for this book is available from the British Library

ISBN: 978-1-032-57429-5 (hbk)
ISBN: 978-1-032-57430-1 (pbk)
ISBN: 978-1-003-43932-5 (ebk)

DOI: 10.4324/9781003439325

Typeset in Sabon
by SPi Technologies India Pvt Ltd (Straive)

CONTENTS

FIGURES

TABLES

ABOUT THE EDITORS

Dr Esmat Zaidan is the Associate Dean for Academic Affairs and Associate Professor at the College of Public Policy, Hamad Bin Khalifa University, Qatar Foundation. She previously held positions at Qatar University, United Arab Emirates University, and the University of Waterloo in Canada, where she earned her master's and PhD degrees. Her previous professional appointments include the World Bank, the Norwegian Agency for Development Cooperation (NORAD), and the United Nations (UNDP). Dr Zaidan's research focuses on sustainable development strategies and policy planning, with an emphasis on human-centred sustainability policies and people-centric cities. She explores the dynamics between technological deployment and human interaction. This includes the policy pathway for energy-efficient buildings, sustainable, resilient, and people-centric urban areas, as well as further study on impacts of technological change on urban life including smart cities and net-zero carbon communities. She particularly focuses on the ways innovations in data-driven modelling and machine learning are shaping sustainability policies, and further positive social, economic, and environmental outcomes. She has published over 65 articles, predominantly in Q1 journals indexed by Scopus and ISI Web of Science, and has been recognised as one of the global 2% Most Cited Scientists in her field.

Dr Imad Antoine Ibrahim is an Assistant Professor in Environmental Law and Governance at Department of Governance and Technology for Sustainability, Faculty of Behavioural, Management and Social Sciences, University of Twente, Enschede, the Netherlands. He is also a Senior Research Associate at gLAWcal – Global Law Initiative for Sustainable

Development, Essex, United Kingdom. He has been working on global issues from a legal perspective for the last decade. He is currently involved in several projects in places such as South Africa, MENA region, Central Asia, and China. In the past, he was a Non-Resident Research Fellow at the Center for Innovation in Gas Research and Utilization (CIGRU) & Institute of Water Security and Science (IWSS), West Virginia University, United States; a visiting researcher at the Environmental Law Institute (ELI), Washington, DC, United States; a Junior Fellow in the Global Research Network Think Tank, United Kingdom; and an Early Stage Researcher at the Dutch Research Institute for Transitions (DRIFT), Rotterdam, the Netherlands. He was also a Resident Research Fellow at the Center for United Nations Constitutional Research – CUNCR, Brussels, Belgium, and an Energy Policy expert at the Lebanese Oil & Gas Initiative – LOGI, Beirut, Lebanon. He worked as an EU Commission Marie Curie Fellow at Tsinghua University School of Law, THCEREL – Center for Environmental, Natural Resources & Energy Law, at the CRAES – Chinese Research Academy on Environmental Sciences in Beijing (China) and at the Business School, Research Center for Climate Change and Trade and at Beijing Normal University. He also worked as a researcher in several European and Italian institutes and universities such as the University Institute of European Studies (IUSE) and Università degli Studi del Piemonte Orientale, Novara. He was a trainee lawyer at Al Jad Law Firm reviewing international conventions related to energy. He holds a PhD in Law from the Institute of Law, Politics and Development (DIRPOLIS), Sant'Anna School of Advanced Studies, Pisa, Italy; a master's degree in European Interdisciplinary Studies, from the College of Europe, Bruges, Belgium, and a bachelor's degree in Law from the Lebanese University "Filiere Francophone de Droit", Beirut, Lebanon.

Dr Elie Azar is a Canada Research Chair and Associate Professor of Civil and Environmental Engineering at Carleton University in Ottawa, ON, Canada. His primary research focuses on integrating research methods and knowledge from different disciplines to promote a sustainable, resilient, and people-centric built environment. Examples of these disciplines include building energy modelling and design, occupant behaviour and comfort modelling, machine learning, and urban energy planning. Dr Azar has led or contributed to research projects exceeding $7.5 million in funding and supervised more than 30 graduate students. He has coauthored more than 150 scientific publications, including four books on artificial intelligence, sustainability, smart cities, and carbon neutrality. He is also frequently invited to contribute to intergovernmental publications, such as UNEP-CCC Climate Technology Progress Reports. Dr Azar has received numerous distinctions and recognitions, such as best paper awards and a feature on the "World's Top 2% Scientists" list prepared by Stanford

University and Elsevier in 2022. Before joining Carleton University, Dr Azar was a faculty member at Khalifa University in the United Arab Emirates. He earned his B.Eng. in Mechanical Engineering from Polytechnique Montreal and his MSc and PhD in Civil and Environmental Engineering from the University of Wisconsin-Madison.

CONTRIBUTORS

Alexander Smit is a PhD candidate at the Centre for Media and Journalism Studies and the Digital Inclusion Lab at Groningen University. His study focuses on the sub-target group of functionally illiterate and/or low-literate adults, and examines how they understand, conceptualise, and practice digital literacies within their everyday lives, what aspects of contemporary digital literacy frameworks they experience as problematic and/or important in their civic and daily activities, and how they experience digital exclusion and inequalities. Through qualitative mixed methods he explores how digitalisation affects civic, social, cultural, political, and personal dimensions of disadvantaged citizens.

Ana Rita Antunes completed a bachelor's degree in mathematics (2017) and a master's in systems engineering (2019) at the University of Minho, Portugal. She completed post-graduation in data analysis for business at Polytechnic Institute of Cávado and Ave, Barcelos, Portugal. Currently, she is pursuing a doctoral programme in industrial and systems engineering at University of Minho. During her professional experience, she was a researcher at the ALGORITMI and 2AI centres and is an invited teaching assistant in applied statistics and data intelligence at University of Minho. Since August 2023, she has been a Data Analyst at Data CoLAB.

António Ferreira Dias completed his master's in electrical and computer engineering with specialisation in Renewable Energy and Energy Efficiency at the Faculty of Engineering of the University of Porto. He has participated in several European and National Research and Development projects, such as the development of Smart Farming, Smart Metering, IoT Solutions for specific applications in the agricultural sector, using his data modelling and design thinking skills. Currently, he is the Head of the Development & Innovation Unit at SGS Portugal and President at DataCoLAB.

Bárbara Patrício completed her bachelor's (2015) and master's (2017) in biochemistry at the Faculty of Sciences, University of Porto, Portugal. Subsequently, she pursued a PhD in translational medicine at the Sant'Anna School of Advanced Studies, in Italy, as part of an European Union's Horizon 2020 Marie Curie Actions Research and Innovation Program (2021). As professional background, she has experience both in academia as a scientific researcher and in industry as a grant writer analyst and business developer. Since November 2022, she has been the Program Manager in Health and Pharma at Data CoLAB.

Brittany Haemi Lee is an urban technologist with nearly a decade of experience spanning the public and private sectors, particularly in real estate, telecommunications, and technology. She specializes in developing go-to-market strategies and has successfully sold enterprise software solutions for municipalities. Currently, she serves as an Account Manager for the Utilities and Municipalities sectors at Esri Canada, where she supports the implementation of geospatial technologies to enhance urban infrastructure and governance. Brittany is the founder of Urban Techies, a nonprofit organization with a global network of over 300 members across eight cities. The organization fosters collaboration among urban professionals dedicated to advancing smart and sustainable city solutions. A recognized thought leader in the smart cities domain, Brittany has delivered keynote presentations and participated in panel discussions at leading international conferences, including the Smart City Expo World Congress, GLOBE Series, and Urban Land Institute, with engagements in Spain, Brazil, Qatar, and Canada. She also contributes to Spacing, a Canadian urbanism magazine, where she explores topics related to urban innovation and technology. Brittany holds a Bachelor of Commerce (Honours) from Queen's University. Raised in Dubai, U.A.E., she currently resides in Toronto, Canada.

Elena Reshetova is Lead Energy Commodities Analyst at Fortum, one of the largest power companies in the Nordics. She is responsible for analysing global macroeconomic trends, oil, coal, gas, and carbon markets as well as their impact on the continental European and Nordic power prices. She provides actionable market intelligence by explaining intraday price volatility, anticipating mid-term risks, and developing long-term scenarios. Elena has researched global natural gas markets, challenges and opportunities in integrating renewables into national energy systems, energy transitions in remote communities of the Arctic and Southeast Asia, and urban energy systems of large ASEAN cities. She has previously worked with government agencies, international organisations, and think tanks in the United States and Singapore. Elena holds a PhD in Public Policy from the National University of Singapore and an MA in International Affairs from Boston University.

Harald Borgebund holds a PhD in political philosophy from the University of York (UK). His research and teaching interests are liberal democracy, democratic theory, political communication, and the role of capitalism and technology in contemporary society.

Henrik Skaug Sætra is a political scientist with a broad and interdisciplinary background and approach, mainly focusing on the political, ethical, and social implications of technology. He focuses specifically on the sustainability-related impacts of AI, and he has published extensively on the relationship between technology and sustainable development and the UN Sustainable Development Goals.

Joëlle Swart is an Assistant Professor at the Centre for Media and Journalism Studies at the University of Groningen. Her research focuses on changing patterns of news use and how users develop practices, norms, and habits around news and journalism in digital environments. Her work has been published in various international journals, including *Digital Journalism*, *New Media & Society and Media*, and *Culture & Society*.

Jungyoon Park is a Senior Researcher at Center for Global Community, University of Myoung-Ji (Seoul, South Korea), where she coordinates various research projects, specialising in Local Planning and Public Administration. Jungyoon Park completed her PhD in Urban Planning at Université Paris-Est Créteil (UPEC) in 2009. Her research interests lie in the area of Sustainable Local development, Smart city and Participatory Planning. Currently, she participates in several research projects supported by the Ministry of Interior and by numerous state-owned enterprises (Korea Rail Network Authority, National Health Insurance Service, Arts Council Korea, etc.) in Korea. Previously, she served as a Research fellow at the Korea-Africa Center (supported by the Ministry of Foreign Affairs in Korea) for two years, organising various research projects for African Urban studies and business meetings for Korea-Africa Business relationship.

Kris Hartley is Assistant Professor of Sustainability and Enterprise at Arizona State University, School of Sustainability. He researches the role of public policy in technology-enabled sustainability transitions, with a focus on circular economy. He has published books with Cambridge University Press and Routledge Press, and peer-reviewed articles on a variety of topics, including environmental policy, smart cities, and global development. Kris has previously been a Fulbright U.S. Scholar (2020) and has held faculty appointments at Cornell University, University of Melbourne, and City University of Hong Kong. He serves on the editorial boards of several journals, including *Journal of Sustainability*, *International Journal of Water Resources Development*, *Australian Journal of Public Administration*, *Policy Design and Practice*, and *Journal of Circular Economy*. Kris holds a PhD in Public Policy from the National University of Singapore (Lee Kuan Yew School of Public Policy), Master of City Planning from the University of California, Berkeley, and Bachelor of Arts in Classics from the University of Tennessee.

Kristina Khutsishvili is a researcher in the Ethics of Artificial Intelligence, Connected Places, and Public Sector Decision-making. She has served as the leader of the work package on Ethical and Inclusive Engagement in Practice for the CommuniCity Horizon Europe project. She holds an interdisciplinary doctorate degree from Scuola Superiore Sant'Anna in Pisa, Italy, and has conducted postdoctoral research at the Informatics Institute of the University of Amsterdam.

Leila Hennaoui is an Associate Professor of International Law at Hassiba Ben Bouali University of Chlef, Algeria. Her expertise lies in public international law, with a prior focus on nuclear non-proliferation. In recent years, her research has broadened to explore the legal implications of technological advancements. This chapter represents her growing interest in the international dimension of smart cities and the crucial role of law in shaping their sustainable and equitable development. Professor Hennaoui actively contributes to the legal community by teaching international law and human rights courses and participating in research projects.

Lynn Abdouni, PhD, is a research scientist at the University of Georgia's College of Engineering. Her work bridges design and geospatial analysis. At the Institute for Resilient Infrastructure Systems, she integrates systems thinking, science communication, and stakeholder engagement to promote the resilience of natural and built environments. Her focus includes generating maps, visualizing data, and advancing actionable strategies for resilience planning.

Marcel Broersma is Full Professor and Director of the Centre for Media and Journalism Studies at the University of Groningen. He is one of the coordinators of the national VSNU Digital Society research programme and Director of the Dutch Research School for Media Studies (RMeS). His research focuses on the interface between the digital transformation of journalism, social media, changing media use, and digital literacy and inclusion. Broersma has received numerous grants for his research, and his work is published widely.

Marco Carlotti is a PhD Candidate in Urban Planning at Sorbonne University (Paris, France). He received a bachelor's degree in Social Sciences from the Catholic University of Paris (2020), a bachelor's degree in Philosophy from Paul-Valéry Montpellier 3 University (2021), as well as a master's degree in Urban Planning from Sorbonne University (2022). His research interest focuses on the concept of "urban ecological sustainability", particularly through the study of eco-cities in France, Morocco, and the United Arab Emirates.

Margarida Prozil completed her master's in biomedical engineering (2010) at the University of Minho, Portugal. She also holds a post-graduate degree in healthcare economics and management (2018) from the Faculty of Economics, University of Porto, Portugal. Throughout her professional journey, she has held various positions in the field of digital transformation and data intelligence at leading technology and healthcare companies. Since September 2023, she has been serving as the Head of Data at Data CoLAB.

Mariana Silva completed her master's in mechanical engineering with a thermal energy specialisation (2018) at the Faculty of Engineering of the University of Porto. Between 2019 and 2021, she was a researcher and a development engineer at the Institute of Science and Innovation in Mechanical Engineering and Industrial Engineering, where she participated in energy recovery research projects. She was also a HVAC Technical Consultant, and she was responsible for Energy Certification of Nonresidential Buildings. Since 2022, Mariana has been a Data Analyst at Data CoLAB, participating in Power BI analysis projects in Portugal.

Maud Rebergen is a PhD candidate at the Centre for Media and Journalism Studies, and a member of its Digital Inclusion Lab, at the University of Groningen. Her work resides at the intersection of public administration, science and technology studies, critical data studies, citizenship studies, sociology, digital literacy, and digital inclusion. In her PhD project, she posits the hypothesis that the widespread adoption of digital technologies in society leads to social exclusion of marginalised populations. She specifically scrutinises how the implementation of digital technologies in Dutch governance has redefined the notion and experience of citizenship.

Øyvind Stokseth is a trained civil engineer and has extensive experience with innovation and digitalisation. He is the Head of smart city initiatives in Halden Municipality but is currently on leave to pursue a PhD at Østfold University College. He has a particular interest in the use and implementation of technology.

Patrizia Ingallina is Professor of Urban Planning and Urban Design Project at Sorbonne University (Paris, France) where she completed her master's degree in Urban Planning. She has 35 years of experience in higher education (master's and PhD) and teaches regularly at Sorbonne University, Abu Dhabi. She obtained her PhD in Geography and Urban Planning from the University of Paris 1 – Panthéon Sorbonne (1993) and her "Habitation à Diriger les Recherches" from the University Paris-Est (2003). She also holds a diploma in architecture from the Faculty of Architecture of the University of Florence (1982), as well as a specialisation in Landscape Architecture (École d'Architecture de Versailles, 1989). Her research and practice are centred on urban design project processes ("projet urbain"). More recently, her research approach is based on an analysis of interpretative models and experiential dimensions of knowledge-based urban design projects involving more and more ecological transition for the protection of urban ecosystems. Professor Ingallina has provided expertise for the Phosboucrâa Foundation (OCP – Office Chérifien des Phosphates) in Morocco. She is also consulted by public and private organisations as an expert urban planner and architect and has been an invited speaker in many universities, research institutes, and international associations (Peking University, University of Seoul, University of Lausanne, Fordham University, etc.).

Ridoan Karim is a lecturer at the Department of Business Law & Taxation, School of Business, Monash University Malaysia. He has designed the Cyber Law Unit at Monash University Malaysia for both undergraduate and postgraduate studies. Dr Ridoan is passionate about the legal and regulatory aspects of emerging technologies and believes that technology's just/fair development is the future of human civilisation. He has worked as a consultant and fellow on various projects funded by the University of Malaya, Monash Data Futures Institute (MDFI), the Malaysian Ministry of Higher Education, and so on. Ridoan Karim holds a Doctor of Philosophy (PhD) from the University of Malaya, Malaysia. He also obtained a Master of Business Administration (MBA) from the University of Chichester, UK; a Master of Comparative Laws (MCL) from International Islamic University Malaysia (IIUM); and a Bachelor of Laws (LLB) from BRAC University, Bangladesh.

Shisong Jiang is an Associate Professor of International Law at Chongqing University Law School. He holds a PhD in Law from the University of Antwerp and a PhD in "Politics, Human Rights, and Sustainability" from the Scuola Superiore Sant'Anna (2020). His research primarily lies at the intersection of international law, urban governance, and global health, with a particular emphasis on the sociology of international law. He has published works examining

Chinese cities' practice of international law and their role in China's legal approach to the Belt and Road Initiative. Currently, he is the principal investigator of a funded research project (2023–2025) focusing on the interaction between cities and international law in the context of global governance competition among major powers.

Xinyue Zheng is an LLM candidate at the School of Law, Xiamen University, with a focus on international taxation law and global governance. She earned her LLB from Chongqing University, where her thesis received the prestigious Best Thesis award for Chongqing Municipality. Zheng has served as an associate researcher on a Chinese National Social Science Foundation project examining taxpayers' right to information and has rich experience in the Philip C. Jessup International Law Moot Court Competition.

FOREWORD

Martin de Jong

Smart city development as a topic of empirical investigation is beginning to reach the age of maturity. This obviously implies that a great many studies have already offered a great variety of angles. There is an ample literature on the engineering aspects of ICT, IoT, technical requirements, standardisation, and many more. A rich literature on the urban planning and development aspects has also emerged, a bit later and partly in parallel to this previous one. Although they should be logically connected and be using findings from the other stream as input for their own analyses and considerations, this is in fact only partially the case because the disciplines they lean on, the concepts they apply, and the analytical frameworks they adopt to derive their findings are very different and at times at odds with one another. The result is an interesting and variegated but also somewhat fragmented and amalgamated object of study. Smart cities: you like them, or you do not. You believe they will help humanity as a whole forward with path-breaking new technologies that expand ever more data-intense horizons with greater opportunities to save energy, monitor harmful emissions, and/or deliver products that consumers crave. Or you assume that a substantial or even growing segment of the population is digitally illiterate and will miss out on those wonderful opportunities, that rebound effects after saving electricity due to better information will just lead people to step up their use afterwards, and that lazy carelessness of online customers drives them to share vital personal information allowing big capital to intensify its further financial exploitation. And frankly, good arguments can be provided for any of these lines of reasoning, although they do seem at least at first sight to lead to incompatible understandings of the situation at hand and potentially contradictory policy advice. Apparently smart cities and what they represent are not as conceptually unambiguous as one might have hoped, and the social implications of their emergence and evolution remain contested and controversial in the end.

What makes the contribution of the authors in this volume special is that they broach three relative new avenues for research that will allow us to view smart city development from an optimistic and constructive angle. The various strengths and weaknesses which smart cities engender are acknowledged, but an effort is made to come to awareness and knowledge and find an opportunity to strengthen the strengths and weaken the weaknesses. These avenues are essentially citizen involvement, political governance, and legal issues in smart cities. In all three areas, the invited authors have collected empirical evidence especially from emerging or developing countries that aim to draw lessons from elsewhere in the world yet face their own specific context-specific challenges. While my own recent thoughts on the degrees of freedom that public organisations and civil society have in going against the pervasive influence of 21st-century data-driven capitalism are rather on the grim side and expect dominant urban growth coalitions simply to have found a new domain for reinvesting and accumulation of financial resources (Jong, Joss and Taeihagh, 2024), it cannot be denied that resting on one's laurels and witnessing the world go down cannot possibly be the answer to a discouraging status quo. We need new thoughts and ideas to find out how bottom-up civil society-driven smart urbanism works, and we need to know how to avoid that crucially important technical standards that become the playball of cynical geopolitics. It is imperative that professional learning programmes are established for familiarising vulnerable groups with the requirements of the digital age, and we can at least strive for smart cities that really cut energy use, reduce ecological footprint, and improve climate resilience. That is far from easy, but nonetheless vital to secure a liveable future. If this book contributes to idea generation on this front and entices a certain number of policy analysts and policy makers to adopt these useful ideas, that would be great. Do we not all, at least in theory, believe that data belong to the people?

Reference

Jong, Martin de, Simon Joss, and Araz Taeihagh (2024). Smart cities as spatial manifestations of 21st century capitalism. *Technological Forecasting and Social Change 202*, 123299.

ABBREVIATIONS

ASToN	African Smart Town Network
APIs	Application Programming Interfaces
AI	Artificial Intelligence
ADS	Automated Decision Systems
CIT	Centro de Inteligência Territoral
COPPA	Children's Online Privacy Protection Act
CCDR	Cities Coalition for Digital Rights
CRM	Customer Relationship Management
DMC	Digital Media City
EID	Electronic Identification
ERP	Enterprise Resource Planning
ESP	EskomSePush
CETO	Ethics Board of the University of Groningen
ECJ	European Court of Justice
ENOLL	European Network of Living Labs
EU	European Union
XR	Extended Reality
ETL	Extract, Transform, and Load
FTC	Federal Trade Commission
GDPR	General Data Protection Regulation
RUED	General Plan of Local Energy
RUEN	General Plan of National Energy
GIS	Geographic Information Systems
GPS	Global Positioning System
GUI	Graphic User Interface
BPTJ	Greater Jakarta Transportation Agency

GHG	GreenHouse Gas
G20	Group of Twente
HIPAA	Health Insurance Portability and Accessibility Act
IT/OT	Informational Technology and Operational Technology
ICT	Information and Communication Technologies
IMD	Institute for Management Development
IESE	Instituto de Estudios Superiores de la Empresa
INE	Instituto Nacional de Estatística
IPMA	Instituto Português do Mar e da Atmosfera
ICP	Integrated Communications Provider
IAP2	International Association for Public Participation
IEC	International Electrotechnical Commission
ISO	International Organization for Standardization
ITU	International Telecommunication Union
IoT	Internet of Things
IP	Internet Protocol
BKSP	Jabodetabek Development Cooperation Agency
KBC	Knowledge-Based Cities
ICLEI	Local Governments for Sustainability
LM6E	Lycée Mohammed VI d'Excellence
MFA	Material Flow Analysis
MAC	Media Access Control
M&E	Media Entertainment
MEMR	Ministry of Energy and Mineral Resources
MHU	Ministry of Housing and Town Planning
NT1	Native Dutch
C-ITS	Next-Generation Intelligent Transport System
NICT	New Information and Communication Technologies
NTP	New Towns Program
NT2	Non-Native
OCP	Office Chérifien des Phosphates
OLS	Object Level Security
OASC	Open and Agile Smart Cities and Communities
OECD	Organization for Economic Cooperation and Development
PDPA	Personal Data Protection Act
PbD	Privacy by Design
PETs	Privacy-Enhancing Technologies
PPPP	Public Private People Partnership
RFID	Radio-Frequency Identification
RLS	Row Level Security
SMG	Seoul Metropolitan Government
SNS	Serviço Nacional de Saúde
SME	Small-to-Medium-sized Enterprises

SADV	Société d'Aménagement et de Développement Vert
SQL	Structured Query Language
SDGs	Sustainable Development Goals
TEF	Testing and Experimentation Facility
VR	Virtual Reality
WeGo	World Smart Sustainable Cities Organization
WTO	World Trade Organization
UCLG	United Cities and Local Governments
UN	United Nations
US	United States
UM6P	University Mohammed VI Polytechnic

INTRODUCTION

Esmat Zaidan, Imad Antoine Ibrahim, and Elie Azar

In the 21st century, urban environments face unprecedented challenges and opportunities. Rapid technological advancements, coupled with the increasing pressures of climate change, population growth, and the need for sustainable development, spurred the emergence of a new urban paradigm – the smart city. Smart cities, powered by technologies like the Internet of Things (IoT), artificial intelligence (AI), and big data, are reshaping how cities operate, interact with citizens, and address complex urban challenges (Harrison et al., 2010; Kitchin, 2014). These cities promise increased efficiency, reduced environmental impact, and improved quality of life, creating a new model for sustainable urban development in the digital age (Caragliu et al., 2011). However, while smart cities hold great potential, they also raise critical questions about privacy, inequality, and governance that must be addressed for these innovations to truly benefit all urban residents.

As we move further into the 21st century, the world is urbanising at an unprecedented rate. According to the United Nations (2018), more than 68% of the global population is projected to live in cities by 2050, which will add immense pressure on urban infrastructure, resources, and services. Urban spaces are becoming denser and more complex, often amplifying existing challenges such as traffic congestion, pollution, inadequate housing, and social inequality (Batty et al., 2012). In response to these issues, the concept of the smart city has emerged as a viable solution to manage urbanisation by leveraging cutting-edge technologies. By integrating IoT sensors, AI algorithms, and real-time data analytics, smart cities aim to optimise urban management, enhance public service delivery, and promote sustainability (Schaffers et al., 2011).

DOI: 10.4324/9781003439325-1

However, while these technologies promise more efficient and sustainable cities, they also introduce new layers of complexity. Privacy is a major concern, as smart cities collect vast amounts of personal and behavioural data through ubiquitous sensors and surveillance systems, raising questions about who controls this data and how it is used (Townsend, 2013). Furthermore, the deployment of AI systems in urban management can perpetuate algorithmic biases that disproportionately affect marginalised communities (Eubanks, 2018). The digital divide is another critical issue – while some citizens may benefit from access to smart technologies, others may be left behind due to lack of digital literacy, access to technology, or economic inequality (Van Dijk, 2020). As smart cities grow, these challenges demand greater attention to ensure that the benefits of urban innovation are distributed equitably.

This book, *Smart Cities to Smart Societies: Moving Beyond Technology*, brings together a range of perspectives and insights from experts across different fields to explore the multifaceted nature of smart cities. Through its chapters, the book highlights the evolution from technology-driven smart cities to more inclusive, citizen-centred smart societies. The term "smart society" represents a shift from focusing solely on technological solutions to recognising the importance of social, cultural, and governance dimensions in building sustainable, equitable, and inclusive urban environments (Sassen, 2010; Albino et al., 2015). In this context, smart cities must move beyond being hubs of innovation to becoming environments where governance, data ethics, and citizen empowerment play a central role.

The rise of smart cities poses a fundamental question: How can we balance the technological efficiency of smart urban environments with the need for inclusivity, privacy, and ethical governance? While advanced technologies like IoT and AI have revolutionised how cities are managed, they also introduce concerns about surveillance, algorithmic fairness, and widening social divides (Kitchin, 2016). If these issues are left unaddressed, the future of smart cities could be one where the benefits of innovation are concentrated in the hands of a privileged few, exacerbating existing inequalities rather than alleviating them (Graham & Marvin, 2001).

This book addresses these dilemmas by rethinking the governance of smart cities through a human-centred lens. The chapters explore how cities can leverage technological advancements while safeguarding citizen rights, ensuring equitable access to resources, and fostering community engagement. In doing so, the book emphasises the importance of not only building smart cities but also ensuring they are just and sustainable. Governance frameworks, legal structures, and community-driven innovation must work in tandem to ensure that smart cities foster environments of fairness, transparency, and citizen participation (Shelton et al., 2015).

The evolution of smart cities

The concept of smart cities has evolved significantly over the past few decades. Initially focused on technology and infrastructure, the notion of the smart city has expanded to include social, economic, and environmental dimensions. Today, smart cities are not just about connectivity and automation but also about enhancing citizens' quality of life, promoting sustainability, and fostering inclusive governance.

This edited volume seeks to capture the complexity of smart cities by offering a holistic examination of the technological, legal, social, and ethical issues that accompany smart urban development. Each chapter in this book provides a unique lens through which to understand the development and governance of smart cities, showcasing global case studies, critical analyses, and forward-thinking solutions.

Key themes and chapter overview

This book, *Smart Cities to Smart Societies: Moving Beyond Technology*, is divided into thematic chapters that explore the core components of smart cities, addressing how technological innovation, governance, and legal frameworks intersect with the need for inclusivity, citizen empowerment, and sustainability. The chapters form a comprehensive narrative, examining how cities can evolve into smart, sustainable, and inclusive societies. By bridging technology with human needs and governance, this book offers insights into the multifaceted development of smart cities across different regions and cultural contexts.

In Chapter 1, Øyvind Stokseth, Harald Borgebund, and Henrik Skaug Sætra argue that while smart cities are often perceived through a technological lens, the preferences, attitudes, and engagement of citizens are paramount to shaping successful urban environments. This chapter explores the models of citizen participation, such as co-creation and participatory design, as crucial for ensuring that smart cities are responsive to the needs of the people who live in them. A bottom-up approach is advocated, where active community involvement is seen as integral to the development of smart cities. This chapter lays the groundwork for the rest of the book, which emphasises that technology alone is not enough – citizen engagement must be at the core of any smart city initiative.

Chapter 2 by Alexander Smit, Maud Rebergen, Joëlle Swart, and Marcel Broersma delves into the critical issue of digital inclusion, particularly for marginalised or disadvantaged groups. Through two case studies, the authors examine how both formal and informal education settings can bridge the digital divide, which is a significant challenge in smart cities. Without digital literacy, many citizens are excluded from the benefits of smart city technologies. This chapter advocates for a holistic approach to digital literacy, integrating technical skills with social capital and community needs. In linking this chapter to the previous one, it becomes clear that inclusion and citizen participation go hand-in-hand, as both are necessary for building truly smart and equitable cities.

In Chapter 3, Kristina Khutsishivili shifts the focus to AI and its potential to empower citizens within smart cities. The chapter goes beyond the idea of AI as a tool for automation and efficiency, proposing that AI can also be leveraged to foster greater citizen participation in decision-making processes. Examples of AI-driven empowerment projects are explored, demonstrating how co-creation between technology providers and residents can ensure equitable outcomes. This chapter builds on the themes of inclusion and participation from the previous chapters, reinforcing the idea that technology, when combined with ethical governance, can empower rather than alienate citizens.

Chapter 4, written by Brittany Haemi Lee, contrasts the top-down and bottom-up approaches to smart city development, using case studies from Morocco and South Korea. While top-down, centralised government planning can lead to rapid technological advancements, it often lacks the local relevance and citizen engagement that bottom-up efforts provide. On the other hand, community-driven approaches result in more innovative, localised solutions that are more responsive to the needs of citizens. This chapter emphasises the need for a balance between these approaches, linking back to earlier discussions on the importance of active community involvement in smart city development. The notion of co-creation, mentioned in Chapters 2 and 4, is further developed here as a key component of the bottom-up approach.

Chapter 5 by Ana Rita Antunes, Mariana Silva, Bárbara Patrício, Margarida Prozil, and António Ferreira Dias addresses the effective use of public data in smart city governance. This chapter explores the challenges associated with data integration, privacy, and transparency, which are crucial to maintaining citizens' trust and fostering equitable development. Case studies like the Territorial Intelligence Center in Portugal exemplify how public data can be used to improve urban management while respecting privacy rights. This chapter links back to the earlier discussions on AI and citizen empowerment (Chapter 4), as data-driven decision-making is integral to creating responsive and efficient cities. However, governance structures must evolve to ensure that data is managed responsibly, which sets the stage for the subsequent discussions on governance in energy transitions (Chapter 7).

In Chapter 6, Elena Reshetova and Kris Hartley focus on energy governance, using Jakarta as a case study for exploring how cities can manage the transition to renewable energy. The authors argue that governance structures must evolve to ensure accountability in energy management, particularly in the context of sustainability and climate change mitigation. This chapter highlights the importance of transparency and citizen engagement in urban energy transitions, linking back to the broader theme of governance explored in Chapter 6. Additionally, this chapter serves as a prelude to the discussion on eco-technological innovation (Chapter 8), where the integration of technology and sustainability is further examined.

Chapter 7 by Patrizia Ingallina, Jungyoon Park, and Marco Carlotti builds on the energy governance discussions in the previous chapter by examining the intersection of technology and ecology in smart city planning. Using case studies from Morocco and South Korea, the authors introduce the concept of "eco-technological cities", where technological innovation is integrated with ecological sustainability. This chapter advocates for a closer alignment between environmental goals and smart city development, reinforcing the idea that cities must evolve beyond mere technological hubs to become models of sustainability. This chapter creates a strong linkage to the platform urbanism discussed in Chapter 8, where technological innovations are applied to informal settings.

In Chapter 8, Lynn Abdouni introduces the concept of platform urbanism, examining how digital platforms like Uber and other gig economy services are shaping urban life in informal or underdeveloped areas. While platforms can provide essential services in cities with limited infrastructure, they also raise concerns about privacy, inequality, and governance. The chapter builds on earlier discussions of inclusion (Chapter 3) and data governance (Chapter 6), advocating for community-driven platform solutions that benefit all citizens, including those in informal settlements. The challenges of governance and equitable development discussed in Chapters 7 and 8 are further explored here, showing how platform-based solutions can either exacerbate or alleviate inequalities in smart cities.

Chapter 9 by Shisong Jiang and Xinyue Zheng takes a geopolitical turn, examining how smart cities are becoming new battlegrounds for global influence. The authors focus on how major powers like China and the United States are using smart city technologies to expand their influence, both domestically and internationally. This chapter raises critical questions about data sovereignty, surveillance, and the global implications of smart city development, linking back to discussions on data privacy (Chapter 6) and platform governance (Chapter 9). The themes of governance and accountability, central to Chapters 6, 7, and 8, are also relevant here as they pertain to international relations and global competition.

In Chapter 10, Leila Hennaoui explores the legal challenges and opportunities that smart cities face on the international stage. The author argues for stronger legal frameworks that can guide smart cities towards sustainability and the achievement of global goals such as the United Nations (UN) Sustainable Development Goals (SDGs). This chapter ties into the broader discussions on governance (Chapters 6, 7, and 8) by emphasising the importance of clear legal standards for ensuring that smart city innovations contribute to global sustainability efforts. The chapter also connects with the geopolitical discussion in Chapter 10, as international legal frameworks are key to managing the global competition over smart city technologies.

In Chapter 11, Ridoan Karim addresses one of the most pressing issues in smart city development – data privacy. The chapter explores the legal and ethical challenges of data collection, particularly in light of frameworks like the General Data Protection Regulation (GDPR). The author advocates for stronger Privacy Enhancing Technologies (PETs) and a human-centric approach to data governance. This chapter ties back to discussions in Chapters 6 and 10 on the role of public data, governance, and international competition, reinforcing the idea that privacy and data security are crucial to building trust in smart cities.

Towards smart societies

The collective insights presented throughout this volume emphasise that the future of smart cities depends not solely on the adoption of cutting-edge technologies but also on their ability to evolve into smart societies – urban environments that prioritise inclusivity, sustainability, and citizen empowerment. The chapters in this book argue that technological advancements, while crucial, must be integrated with human-centric governance, ethical data management, and inclusive urban planning to ensure that smart cities truly serve the needs of all residents, rather than merely a select few.

The concept of smart societies represents a paradigm shift from the early technocentric vision of smart cities, where the primary focus was on infrastructure, automation, and efficiency (Hollands, 2008). While IoT, big data, and AI have enabled unprecedented levels of connectivity and real-time urban management, these technologies alone cannot solve complex urban challenges such as inequality, social exclusion, and environmental degradation (Vanolo, 2014). The true potential of smart cities, as explored in the chapters, lies in their ability to enhance the quality of life for all citizens by fostering participatory governance, ethical technology use, and sustainable development practices (Nam & Pardo, 2011).

This volume highlights the critical role that inclusive governance plays in shaping smart societies. Several chapters discuss the importance of bottom-up approaches, where citizens are not just passive recipients of technology but active contributors to the design and implementation of smart city projects (Caragliu et al., 2011). As the authors argue, for smart cities to be truly sustainable and equitable, they must incorporate mechanisms for citizen engagement, ensuring that the voices of marginalised and disadvantaged groups are heard and accounted for in decision-making processes (Sadowski & Bendor, 2019). This participatory model aligns with the broader shift towards urban governance that emphasises transparency, accountability, and shared responsibility between governments, the private sector, and civil society (Kitchin, 2015).

Sustainability, as explored in Chapters 7 and 8, is another cornerstone of the smart society vision. Technological innovations, particularly in energy management, transportation, and waste reduction, offer significant opportunities for cities to reduce their environmental impact and combat climate change (Hodson & Marvin, 2010). However, as the book demonstrates, sustainability in smart cities must go beyond technological fixes; it requires a holistic approach that integrates green infrastructure, ecological planning, and long-term environmental stewardship (Joss et al., 2019). This approach ensures that cities not only meet the needs of the present but also safeguard the resources and opportunities for future generations, aligning with the UN SDGs (United Nations, 2015).

The legal and regulatory frameworks discussed in Chapters 10, 11, and 12 further underscore the importance of governance in the development of smart societies. As the digital infrastructure of smart cities grows, so do the challenges related to data privacy, security, and sovereignty (Zuboff, 2015). The book advocates for the implementation of robust PETs and the development of international legal standards to protect citizens' rights in an increasingly data-driven world (Jiang & Zheng, 2021). These frameworks are essential for building trust between citizens and governments, ensuring that the benefits of smart city technologies are equitably distributed and do not exacerbate existing inequalities (Greenfield, 2013).

By linking technological innovation with governance structures, ethical considerations, and sustainability, this volume provides a comprehensive blueprint for how cities can evolve into smart societies – urban environments where technology enhances, rather than undermines, human well-being (Calzada & Cobo, 2015). The chapters collectively emphasise that while the tools of the smart city are essential for addressing the challenges of urbanisation, it is the inclusive, citizen-centred approach to governance and planning that will ultimately determine their success. As cities around the world continue to embrace smart technologies, this volume offers crucial insights into how they can ensure these innovations contribute to a more just, sustainable, and equitable future for all.

References

Albino, V., Berardi, U., & Dangelico, R. M. (2015). Smart cities: Definitions, dimensions, performance, and initiatives. *Journal of Urban Technology*, *22*(1), 3–21.

Batty, M., Axhausen, K. W., Giannotti, F., Pozdnoukhov, A., Bazzani, A., Wachowicz, M.,... & Portugali, Y. (2012). Smart cities of the future. *European Physical Journal Special Topics*, *214*(1), 481–518.

Calzada, I., & Cobo, C. (2015). Unplugging: Deconstructing the smart city. *Journal of Urban Technology*, *22*(1), 23–43.

Caragliu, A., Del Bo, C., & Nijkamp, P. (2011). Smart cities in Europe. *Journal of Urban Technology*, *18*(2), 65–82.

Eubanks, V. (2018). *Automating inequality: How high-tech tools profile, police, and punish the poor*. St. Martin's Press.

Graham, S., & Marvin, S. (2001). *Splintering urbanism: Networked infrastructures, technological mobilities and the urban condition*. Routledge.

Greenfield, A. (2013). *Against the smart city*. Do Projects.

Harrison, C., Eckman, B., Hamilton, R., Hartswick, P., Kalagnanam, J., Paraszczak, J., & Williams, P. (2010). Foundations for smarter cities. *IBM Journal of Research and Development*, *54*(4), 1–16.

Hollands, R. G. (2008). Will the real smart city please stand up? *City*, *12*(3), 303–320.

Hodson, M., & Marvin, S. (2010). *World cities and climate change: Producing urban ecological security*. McGraw-Hill Education.

Jiang, S., & Zheng, X. (2021). Smart cities as new battlegrounds in the geopolitics of international law. In *Smart cities to smart societies* (pp. 185–198). Routledge.

Joss, S., Sengers, F., Schraven, D., Caprotti, F., & Dayot, Y. (2019). The smart city as global discourse: Storylines and critical junctures across 27 cities. *Journal of Urban Technology*, *26*(1), 3–34.

Kitchin, R. (2014). *The data revolution: Big data, open data, data infrastructures and their consequences*. SAGE Publications.

Kitchin, R. (2015). Making sense of smart cities: Addressing present shortcomings. *Cambridge Journal of Regions, Economy and Society*, *8*(1), 131–136.

Kitchin, R. (2016). The ethics of smart cities and urban science. *Philosophical Transactions of the Royal Society A: Mathematical, Physical and Engineering Sciences*, *374*(2083), 20160115.

Nam, T., & Pardo, T. A. (2011). Conceptualizing smart city with dimensions of technology, people, and institutions. *Proceedings of the 12th annual international digital government research conference*.

Sadowski, J., & Bendor, R. (2019). Selling smartness: Corporate narratives and the smart city as a sociotechnical imaginary. *Science, Technology, & Human Values*, *44*(3), 540–563.

Sassen, S. (2010). *Cities in a world economy*. SAGE Publications.

Schaffers, H., Komninos, N., Pallot, M., Trousse, B., Nilsson, M., & Oliveira, A. (2011). Smart cities and the future internet: Towards cooperation frameworks for open innovation. In *The future internet assembly* (pp. 431–446). Springer.

Shelton, T., Zook, M., & Wiig, A. (2015). The 'actually existing smart city'. *Cambridge Journal of Regions, Economy and Society*, *8*(1), 13–25.

Townsend, A. M. (2013). *Smart cities: Big data, civic hackers, and the quest for a new Utopia*. W.W. Norton & Company.

United Nations. (2015). *Transforming our world: The 2030 agenda for sustainable development*. United Nations.

United Nations. (2018). *World urbanization prospects: The 2018 revision*. United Nations Population Division.

Van Dijk, J. (2020). *The digital divide*. Polity Press.

Vanolo, A. (2014). Smartmentality: The smart city as disciplinary strategy. *Urban Studies*, *51*(5), 883–898.

Zuboff, S. (2015). Big other: Surveillance capitalism and the prospects of an information civilization. *Journal of Information Technology*, *30*(1), 75–89.

PART I

Smart cities and citizens

1

FOR US, BY US?

The role of citizens in smart cities

Øyvind Stokseth, Harald Borgebund, and Henrik Skaug Sætra

Introduction

Smart city initiatives are heralded as groundbreaking solutions to urban challenges, aimed at enhancing economic efficiency, service quality, and overall quality of life. These initiatives leverage cutting-edge technologies to optimize public services and infrastructure, promising significant benefits for municipalities, businesses, and residents alike (Gracias et al., 2023; Griffy-Brown et al., 2018). While the term *smart city* is often used, we are here usually referring to smart cities *and communities* more broadly, as many of the solutions considered integral to a smart city are also applicable to smaller communities without city status. Furthermore, by discussing smart cities and communities, we also align with Sustainable Development Goal (SDG) 11, which refers to *sustainable* cities and communities (UN, 2015).

However, the vision of smarter cities is not without its critics. Concerns about privacy, surveillance, and the creation of digital divides have sparked a debate on the implications of increasingly digital infrastructure (Wadhwa, 2015; Nordrum, 2023). As digital technologies become more embedded in urban settings, they bring not only opportunities but also challenges that require careful management and ethical considerations.

This chapter focuses on a critical yet often overlooked group of stakeholders: the citizens. While smart cities are usually discussed in terms of technological and economic metrics, we aim to explore the social dimensions that influence and are influenced by smart city technologies. We begin by examining different conceptualizations of a 'smart city,' acknowledging that the term encompasses a broad and sometimes ambiguous range of definitions and expectations (Syed et al., 2021; Helbing et al., 2021). This exploration

DOI: 10.4324/9781003439325-3

serves as a foundation for our stakeholder mapping, which illustrates that citizens, despite being crucial components of any city, often have limited influence over how these technologies are implemented and governed.

We will also explore the various roles citizens could play in the planning, development, evaluation, and management of smart cities. The discussion is framed around the idea that smart city development should not merely involve citizens as passive end users but engage them actively as co-creators of their urban environments. The chapter concludes with a discussion of the essential factors to consider in ensuring that citizens are not just living in smart cities but are also meaningfully engaged in making their cities smarter – based on their own goals and definitions of smartness.

Conceptualizing the smart city

Defining a smart city is challenging, as cities often claim the 'smart' label based on a wide range of criteria. Examples include the implementation of innovative e-governance schemes, the establishment of social learning ventures and community engagement programs, and a focus on sustainable living (Syed et al., 2021). The lack of a clear definition of smart cities poses challenges for policymakers, urban planners, and stakeholders in developing and implementing smart city strategies (Gracias et al., 2023; Griffy-Brown et al., 2018). To effectively determine the most suitable technology for urban use and how it can be intelligently applied, it's crucial to consider the intended users and the objectives of the solution.

Cities, communities, and smart technology

A city is often associated with a large and densely populated urban area, characterized by its significant size, economic activity, and cultural importance. But according to Pile (1999), there is no simple, precise answer to what a city really is. He argues cities typically feature a complex infrastructure system, including transportation networks, communication systems, utilities, and buildings. Cities are often centers for business, trade, industry, education, and government. They play a crucial role in the social, cultural, and economic development of a country or region.

Cities are distinguished from smaller towns and villages by their larger populations, more developed infrastructure, and greater diversity in terms of jobs, educational institutions, healthcare facilities, and cultural offerings. They are usually hubs of innovation and creativity, drawing people from various backgrounds and cultures. This diversity can lead to a rich cultural life but also to challenges such as congestion, pollution, and social inequality. The term smart city normally refers to cities, but it could also be used to refer to smaller towns and communities, as we have mentioned.

The academic debate on smart cities showcases a variety of perspectives related to their desirability and the implications of smart technologies for citizens and other stakeholders. Helbing et al. (2021) explain how some scholars view smart cities as viable solutions to urbanization, sustainable development, and enhanced socio-economic well-being (De Jong et al., 2015). In contrast, others view the concept as leading to an ambiguous and potentially dystopian society (Vanolo, 2016). However, many definitions of smart city technology refer to the integration of various forms of digital and information technologies into urban environments to enhance the efficiency, effectiveness, and quality of urban services, and to address urban challenges. The European Commission characterizes smart cities as urban areas that optimize traditional networks and services using digital technologies, with the objective of enriching the quality of life for residents and bolstering business efficiency (Commission, 2023). Using this definition, we see that a smart city refers to how a place uses technology regardless of whether it is a city, town, or village. Kitchin (2022) explains how developments in smart cities occur within a variety of local settings, each influencing the way these cities are conceptualized and developed and how smart cities can be adapted or utilized to support various objectives or purposes.

Given the various approaches to what a smart city is, discussing the societal consequences of this concept can be challenging. Therefore, we prefer to focus on the societal consequences of implementing smart city technology, as this frees us from being tied to any specific definition of a smart city. To clarify, the term 'smart city technology' refers in this text to the incorporation of information and communication technologies into urban infrastructure. This integration can include things like digital transportation systems, digital technology used to improve efficient energy usage, digital services for public administration, and technologies aimed at enhancing public safety (Syed et al., 2021). The aims are assumed to be, for example, to make cities more responsive to the needs of residents, to reduce environmental impact, to improve the cost-effectiveness of the provision of public services, and to promote opportunities for innovation and economic growth (Gracias et al., 2023).

Smart city solutions are being deployed in cities and communities across the globe, necessitating the need to consider varieties in cultural, political, and economic factors. However, for a city or community to become 'smart' in the sense described here, a number of preconditions related to energy and computing infrastructure must be satisfied. Various nations have embraced a range of smart city solutions, including intelligent transportation systems, security cameras, air quality sensors, and smart lighting. These solutions are tailored to address the specific requirements of each community (Gracias et al., 2023).

Evaluating the implications of smart city technology

Various studies offer differing perspectives on the social impact of smart city technology and the extent to which residents are involved in the development of smart cities. Hollands (2014), for example, considers the social consequences of technological advancements and criticizes the limited role of citizens in discussions of smart cities. Similarly, Boon et al. (2017) discuss the ambiguous roles of citizens and highlight the tendency for corporate-controlled smart cities to foster dependent citizens, while also shedding light on the potential challenges in building citizen-centric smart cities. These varying viewpoints are to be expected, especially considering the absence of a universally accepted definition of a smart city and a precise understanding of the objectives for employing smart city technology. In a city that utilizes smart city technology to collect personal data for control purposes, the social consequences will be markedly different compared to a city that primarily focuses on smart city solutions designed to provide residents with information about weather and local environmental conditions.

The question concerning the role of technocracy and democracy is particularly pertinent here. While technocracy suggests that those with expertise ought to govern to ensure competent decisions, we must also question who sets the agenda and who has the final word in the implementation of smart city technology (Sætra, 2020). This leads to a broader inquiry about the balance between expert decision-making and democratic governance. Such a balance is crucial to avoid scenarios where the deployment of technology could lead to decisions that may not align with the broader public interest or democratic values. Transparency and openness in the involvement process are essential to prevent public skepticism and to ensure that any technological implementation is viewed as legitimate and beneficial. However, we cannot stop at a consideration of the stated purpose of implementing smart city solutions, as (a) the stated intentions might be deceptive and (b) even technologies implemented for benign purposes can now or in the future be used for, for example, control and surveillance purposes (Sætra, 2019).

Moving from theoretical considerations to more quantitative approaches, we find several indexes and rankings used to evaluate and list smart cities based on various criteria like technology integration, sustainability, citizen engagement, and governance. The Instituto de Estudios Superiores de la Empresa (IESE) Cities in Motion Index made by IESE Business School, which is part of the University of Navarra in Spain (IESE, 2024), ranks cities on based dimensions such as technology, environment, social cohesion, mobility and transportation, governance, urban planning, international projection, economy, and human capital. The Smart City Index (Smartcities.org, 2022) ranks cities based on economic and technological data as well as by their citizens' perceptions of how 'smart' their cities are. The International Institute for Management Development (IMD) (2024) Smart City Index developed by the IMD ranks cities based on economic and technological aspects on the one hand, and 'humane dimensions' (quality of life, environment, inclusiveness) on the other. A key question to ask is if smart cities with high scores are good cities for the residents. To answer this question, we must first define what a good city *is*, of course. That is a question beyond the scope of this chapter, but as we show below, we argue that citizens should at least be given a voice when seeking an answer to the question.

Stakeholders and incentives in smart city development

In a smart city, there are many groups with different goals for using technology either explicitly for city development or in contexts where the use of their technology accidentally shapes and influences a variety of stakeholders in the city. While we like to think of the city's residents as key stakeholders, other groups of stakeholders often have more say in how smart city technology is used. How much residents get involved depends on the city's political system, the goals of the technology, the balance of power among different groups, the available opportunities for getting involved, and the willingness and interest of getting involved. However, we must stress that 'citizens' is not a homogenous group with shared interests and preferences. Citizens do not all want the same things. For example, many likely focus on how technology saves money, increases efficiency of public services, creates growth and job opportunities, helps mitigate greenhouse gas emissions. In sum – good things (Vanolo, 2016). However, others are worried about technology that could be used to monitor and control them and want to ensure that technology doesn't leave anyone behind (Vanolo, 2016; Mutambik, 2023).

However, there are also other major stakeholders relevant for understanding the development and management of smart cities and communities. One relevant framework for identifying the key stakeholders is the Quadruple Helix Model often used in innovation studies (see Figure 1.1).

This model expands on the traditional Triple Helix Model, which mainly focuses on three groups of stakeholders as essential for innovation: academia, industry, and the government. The Quadruple Helix adds a fourth dimension: the civil society, or the public. The Quadruple Helix Model in smart city development stands out for its inclusivity and holistic approach, ensuring that urban innovations are not only technologically advanced but also socially relevant and responsive to the needs of all stakeholders, especially the public.

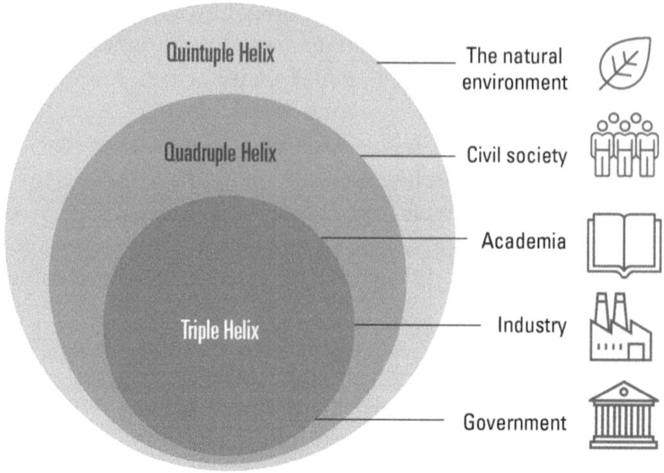

FIGURE 1.1 The three helixes and the relevant actors.

(*Source:* Authors' own)

The four groups of stakeholders have different roles and interests in smart city development and management, as summarized in the following:

1 Civil society (including citizens): Represents the needs, values, and perspectives of the community, ensuring that smart city developments are user-centric and address real-world challenges.
2 Government: Creates policies, regulations, and provides funding, ensuring that smart city initiatives align with public interests and legal frameworks.
3 Industry: Offers technological solutions, investments, and practical applications of innovations that can be utilized in smart city initiatives.
4 Academia: Provides research, innovation, and educational resources, contributing to the development of new technologies and methodologies.

Some have developed the *helix* further and now speak of a *Quintuple* Helix model, which includes a fifth dimension: the natural environment. This expansion encourages stakeholders to consider the ecological impact of their initiatives and to develop solutions that are not only technologically advanced and socially beneficial but also environmentally sustainable (Carayannis & Campbell, 2010). The definition and the purpose of the smart city concept will define if the Quintuple Helix model or Quadruple Helix Model is most suitable (Carayannis & Campbell, 2010). If environmental sustainability is a core component of the smart city initiative, the Quintuple Helix model would be more appropriate as it integrates ecological perspectives (Carayannis & Rakhmatullin, 2014). In contrast, if the focus is more on the interplay between government, industry, academia, and civil society without a specific emphasis on environmental aspects, the Quadruple Helix Model may be sufficient. With our emphasis on stakeholder interests and agency, the natural environment is somewhat peripheral to these concerns, and we'll also argue that all stakeholders can and will have a certain interest in accounting for both the availability of natural capital and the preservation of the natural environment. We will consequently focus on the four groups of the Quadruple Helix Model.

The Quadruple Helix Model recognizes that effective, sustainable smart city initiatives require not just technological solutions but also input from, and collaboration among, diverse stakeholders, including the community that will be directly affected by such initiatives (Arnkil et al., 2010). This model is seen to foster innovation, address societal challenges, and ensure that smart city projects are socially, economically, and environmentally sustainable (Carayannis et al., 2012).

For the various interest groups to jointly contribute ideas and collectively participate in the technology development in cities and urban areas, physical or digital meeting places between the different interest groups must be established. The Quadruple Helix Model, while fostering collaboration among government, industry, academia, and civil society, faces challenges in effectively harmonizing the diverse perspectives and interests of these stakeholders (Carayannis & Rakhmatullin, 2014). The model illustrates the dynamic interplay between university, industry, government, and civil society, (Ivanova & Leydesdorff, 2014), however balancing power among these stakeholders is essential for equitable smart city development. Additionally, aligning the varying goals and communication styles of these diverse groups can be complex, potentially leading to conflicts and inefficiencies in decision-making processes.

Before we move to the discussion of what constitutes proper citizen participation in smart city innovation processes, we will briefly highlight some of the interests and challenges that categorize the different stakeholder groups.

Starting with governmental actors and local authorities, these have a clear interest in promoting growth and innovation and improving public services. Such improvements could entail improved *quality* of the services provided, such as dynamic public transportation routes and services more responsive to the actual and changing requirements of the citizens, or improved cost efficiency of similar-level quality services that allow the municipality to provide more of the same services or to increase efforts in other fields (Nam & Pardo, 2011b). National authorities could of course also see smart city solutions as a way for municipalities to do the same with smaller budgets, or to do more with the same budgets. While we often think of citizens as the primary recipients of public services, industry actors also consume these services and are important contributors to them. This means that local authorities will also have strong incentives to ensure that local industries are competitive and thriving, both to increase the economic activity in the city or community in order to increase their budgets, but also because increased economic activity equals, for example, more jobs and opportunities for citizens. The development of smart cities often involves multiple stakeholders, ensuring that these stakeholders engage in meaningful dialogues is crucial for the successful governance of these initiatives (Hollands, 2020).

The industrial sector, ranging from startups to multinational corporations, primarily seeks to leverage smart city technologies for enhancing competitiveness and operational efficiency (Porter & Heppelmann, 2014). The industry actors will have a keen interest in utilizing technology as much as possible as a tool for addressing challenges that arise in cities. The industry benefits from having close collaboration with local authorities to showcase their products and solutions. However, solutions that are suitable in one city may be less appropriate in another. It can be challenging for local authorities to find a balance between needs and opportunities, and this may result in some cities purchasing solutions that may not be suitable. Sensor technology or citizen platforms that are appropriate in one city may not necessarily be as fitting in another, but this is generally of less concern to the industry (Batty et al., 2012).

Academic and research entities rely on research funding and are therefore interested in initiating projects, often pilot studies, that hold research potential. These projects can be of great value to citizens, but there is also a challenge in that such projects can often be too theoretical, providing limited direct value to residents. Challenges for academia also include maintaining interdisciplinary collaborations that are often required for comprehensive smart city solutions (Leydesdorff & Etzkowitz, 2001).

While a consensus exists among the majority that digital exclusion is an unfavorable outcome, a pertinent question arises: to what degree should the apprehensions of a few hinder the implementation of technology that offers substantial benefits to the broader populace? Furthermore, the degree of citizen engagement is not uniform. Despite the availability of various channels for participation, including physical meetings, assorted surveys, and digital platforms, a considerable number of residents refrain from actively contributing their perspectives, suggestions, and beliefs, despite holding definitive opinions on these matters.

Smart city technology providers typically aim for widespread adoption of their solutions (Komninos, 2014), while academic institutions usually pursue projects that offer potential for research funding (Leydesdorff & Etzkowitz, 2001). Meanwhile, residents generally prioritize solutions that enhance their personal well-being. It can be challenging for the government to balance various inputs, desires, and needs (Nam & Pardo, 2011a).

Discussion: Taking citizens seriously in smart city development

In nations governed democratically, there is a common consensus that cities, whether they are labeled as smart cities or otherwise, ought to be developed in line with the desires of their inhabitants. In Norway, an example of a democratic country, there are certain cases where citizen involvement is mandated by law. In most cases, it is not legally required, but municipalities may choose to implement citizen involvement because such processes capture important information and knowledge (KS, 2023).

The principle of citizen participation and engagement, defined by the International Association for Public Participation (IAP2) as the involvement of individuals in decision-making processes that affect their lives, is central to ensuring that policies align with citizen preferences. This principle is foundational to the implementation of digital municipal services and smart city innovations, although the methods for involving citizens can vary greatly (Gracias et al., 2023). The practical application of this principle, however, can be complex and is not always advisable. In situations that are very technical or urgent, we might need to change the usual ways we involve citizens in managing smart cities to keep things efficient and effective (Nam & Pardo, 2011b). Additionally, the feasibility of involving citizens may be limited by financial and resource constraints, particularly in decisions that have minimal impact on the community or in contexts where public participation could deepen social divides or incite conflict. In such cases, alternative governance approaches that still uphold democratic principles might prove more effective.

Smart city planning must consider the diverse levels of citizen engagement, which range from non-participation to full citizen control (Arnstein, 1969), acknowledging the varying preferences across the community. This entails considerations related to, for example, to what degree citizens can be reasonably expected to see the complex relationships between, for example, innovation and local technology development and the opportunities and services they get as citizens (Fung, 2006). Such complexities are not necessarily reasons for paternalistic and technocratic solutions bypassing citizens; however, they do require tailored decision-making processes and information and educational activities to ensure that the various stakeholders can meaningfully engage with the issues at hand.

Given the variety of definitions surrounding the concept of smart cities, our focus is on examining residents' participation specifically in the implementation of smart city solutions, rather than broadly discussing their involvement in smart city development (Hollands, 2020). Furthermore, we contend that the type of resident involvement required varies based on the specific smart city solution being considered for implementation.

There are several models for resident participation – each suited for different contexts and issues (Arnstein, 1969). Sometimes the goal is to provide information to the residents. This primarily applies to scenarios where residents seek information about a smart city solution either in the process of being implemented or already in place. In these cases, the residents' perspectives may not be relevant for deciding the introduction or application of such solutions. This is particularly true for highly complex technical systems that neither offer insights into nor influence residents' behaviors. These systems exemplify situations where resident input may not effectively contribute to decision-making processes. In other cases, residents should participate in decision-making processes and provide input along the way. In such scenarios, the perspectives and contributions of residents should become vital in determining the adoption and application of these solutions (Innes & Booher, 2004). Engaging residents in this process ensures that their needs and preferences are considered, leading to more effective and inclusive urban development. The ideas of co-production and its closely linked counterpart, co-creation, are frequently mentioned in discussions about smart cities as they pertain to the active participation of citizens in efforts to enhance urban intelligence.

In the context of citizen involvement in smart cities the co-creation and participatory design refer to collaborative processes where citizens actively contribute to the development and design of services, policies, or projects (Sanders & Stappers, 2008). These approaches involve citizens not just as end users or beneficiaries, but as partners in the creative process, ensuring that their needs, ideas, and experiences shape the outcomes. This collaborative method enhances the relevance and effectiveness of solutions by leveraging diverse perspectives and expertise from the community.

Nonetheless, there is a scarcity of documented instances in academic articles that detail practical applications of co-production and co-creation in the development of smart cities. Even rarer are studies examining the effects of these co-production activities on the dynamics of power between citizens and other stakeholders in the urban setting (Romanelli, 2019).

If we return to the Quadruple Helix Model, we see that the roles and impacts of the four actors in the Quadruple Helix Model are as follows:

- Government: Provides regulatory frameworks, funding, and policy support.
- Industry: Drives technological innovation and commercial application.
- Academia: Offers research, expertise, and innovative ideas.
- Civil society (public): Their involvement ensures that smart city initiatives are user-centric, socially inclusive, and reflective of the community's needs and values.

However, the impact of each actor in the Quadruple Helix Model is not static but varies depending on the unique characteristics of each smart city project, including its goals, local context, resource availability, stakeholder dynamics, technological aspects, regulatory environment, and public perception (Carayannis & Campbell, 2009). This variability necessitates a flexible and adaptive approach to stakeholder collaboration in smart city initiatives.

A key question to ask is if, when, and why citizens should not be involved when implementing smart city technology. One way to answer this question is to use either a participatory approach or a paternalistic approach.

Implementation of smart city strategies raises some important questions about the relationship between smart city and democracy, particularly concerning the role of democracy in smart city projects. As smart city approaches often emphasize citizen involvement and trust, it's critical to consider whether involvement is genuinely aimed at gathering diverse opinions or if it's used as an instrument to arrive at predetermined 'correct answers,' thereby influencing the democratic process. This aligns with discussions on deliberative democracy, where the risk of involvement processes being hijacked by powerful groups is a notable concern. These groups, whether they are public servants, politicians, or technocrats, might use citizen participation as a means to justify and legitimize the implementation of smart city technologies under the guise of community benefit.

A technocratic and paternalistic approach in governance implies authority figures making decisions for citizens, often limiting individual choices under the notion of greater good or protection (Dahl, 2008). It entails passive citizen involvement, with authorities leading decision-making. Conversely, a participatory approach emphasizes that everyone's voice is heard and considered. Here, citizens actively participate in governance, emphasizing self-governance and democratic engagement (Habermas, 1985). While paternalism focuses on guidance and control, liberalism champions individual autonomy and informed personal decision-making.

The preceding concerns are visible in the daily activities of municipalities, and we will briefly discuss the case of Halden, in which one of the authors works with smart city development and management. Halden is a smaller Norwegian city with approximately 31,000 residents, and it is enthusiastic about embracing innovation and technological solutions. Halden municipality was awarded by the international analysis agency Analysis Mason as the best Nordic medium-size municipality in the Nordic Digital Municipality Index 2020 (Mason, 2020). The city's research institutions are intent on keeping Halden at the forefront of data collection and sharing. They aspire for the municipality to lead in establishing data sharing infrastructure. Numerous companies have approached the municipality to assist in achieving this goal.

However, Halden faces a significant challenge: its budget is extremely tight, leaving minimal scope for innovation. This raises a crucial question: can the residents contribute their views on prioritizing investment in data sharing infrastructure over other vital sectors like education or healthcare? Furthermore, do the residents grasp the potential benefits of such an investment, and if so, how can their involvement be effectively facilitated? This example shows that a paternalistic approach could be a good choice because of the complexity of smart city solutions. However, it is important that the citizens are informed about the planned use of collected data from the city.

Imagine a different scenario: a local government plans to install a high-tech system in homes with fireplaces. This system involves placing sensors in chimneys to indicate when cleaning is needed. The aim is to move from a fixed cleaning schedule to one based on actual need, which could save money and reduce carbon emissions by decreasing the frequency of visits. The municipality informed residents about installing these sensors, but their input was sought only after all the decisions were made and the plan was ready to go. When it was time to fit these sensors in homes, many residents strongly objected. This leads to a question: would involving residents earlier in the planning process have lessened the protests?

However, to foster a more equitable development of smart cities, we believe it is crucial for public authorities to facilitate the inclusion of academia and civil society by providing digital platforms designed for engagement. Municipalities can establish digital forums and consultation portals that enable these groups to contribute effectively through the introduction of digital citizen services, to urban planning and policymaking. Academic institutions can use these platforms to disseminate research findings that directly influence strategies for urban development, ensuring that technological and scientific perspectives are integrated during the introduction of digital services and planning processes. Civil society, on the other hand, can use these platforms to organize, gather support, and ensure diverse community voices are heard, especially those that are often marginalized in physical engagement settings. When municipalities use such digital platforms, these stakeholders can maintain a sustained influence over the agenda and priorities of smart city initiatives. This approach ensures that the development of smart cities is a collaborative effort that respects and incorporates a wide array of stakeholder insights and needs.

In Helsinki, Finland, the participatory budgeting initiative 'OmaStadi' exemplifies effective citizen engagement in urban planning. From 2019, OmaStadi allocates approximately €8.8 million annually for projects proposed and voted on by residents, ranging from improvements in local parks to community-driven cultural events (Helsinki, 2024). This initiative empowers citizens by directly involving them in budgeting decisions, ensuring that their input shapes the city's development projects, thus reflecting a practical application of civic involvement within the Nordic model of participatory governance.

In this context, a participatory approach seems fitting, as the smart city solution directly involves residents through the installation of sensors in their homes. But how should their involvement be handled (Arnstein, 1969)? Should residents have a say in whether the solution is implemented, or should they just be informed about the installation process and data collection methods? Or perhaps they should have a role in how the gathered data is managed and used? If residents are merely informed and not actively involved, there's a high risk of resistance. Some might worry about increased costs for frequent fireplace users or feel uneasy about their wood-burning habits being monitored. Allowing residents to have a say in the solution's implementation could lead to environmentally beneficial decisions being overlooked in favor of personal interests. A more effective approach might be to invite residents to share their views on data collection and usage. This could reduce potential backlash. Feedback could be collected via a digital platform for residents or through a representative group. It's also important to ensure that all residents receive detailed information about the solution once it's decided upon.

Furthermore, as we consider the role of democracy in smart city projects, it becomes evident that both experts and politicians must work closely together to ensure that technological advancements do not bypass the democratic process. Residents have a legitimate interest in knowing who is responsible for the decisions made regarding SC technology implementation. It is easier for the public to hold politicians accountable at the ballot box, hence the importance of clear and accountable governance structures in smart city initiatives.

In the journey of developing smart cities, citizen involvement is key and can take various effective forms. One popular method is through public consultations, where residents actively give their feedback on proposed projects, either in person during meetings or through surveys (Innes & Booher, 2004). This approach is particularly useful in projects related to urban mobility or construction, such as deciding the design of a swimming pool or determining the best locations for electric scooters.

Another dynamic method is participatory design, inviting citizens to be hands-on in shaping and designing urban solutions. This method can play a crucial role in ensuring that the solutions developed are responsive to the needs and inputs of the community (Sanders & Stappers, 2008). Alongside this, digital platforms provide a virtual space for ongoing interaction, enabling residents to express their ideas and opinions from the comfort of their homes.

Further deepening this collaboration are co-creation workshops and focus groups, which bring together citizens and planners to brainstorm and innovate together. Take, for instance, the implementation of a chimney sensor system, where resident involvement could prove invaluable in fine-tuning the solution to meet community needs. Similarly, the installation of smart water meters presents another opportunity for citizen input. These meters, typically used for accurate water consumption billing, can be adapted for additional functions like alerting residents to unusual water usage or potential leaks, based on their preferences and feedback.

Moreover, initiating pilot projects in specific neighborhoods allows local residents to be directly involved in testing and improving smart city technologies, ensuring these innovations align with their real-life needs and expectations. Accessibility and inclusion are important in digital initiatives (Jaeger, 2011) and can be effectively applied to pilot projects in smart cities to enhance community engagement and ownership.

Different methods of citizen involvement should be chosen based on the needs, opportunities, and desires for input and information sharing. When implementing smart city solutions that will be directly used by residents, it's crucial to ensure inputs that guarantee universal design and prevent digital exclusion (Jaeger, 2011). An example of such a solution could be an app with an associated electric car charger. However, in the above example involving data sharing, it's probably not necessary to place significant emphasis on universal design or digital exclusion.

It is essential that the type of smart city technology being considered for implementation dictates how residents should be involved.

When involving residents, it's crucial to consider how their input will be managed and evaluated. Residents are not a homogeneous group, and conflicting arguments or suggestions may arise. However, it's important that feedback is provided to all residents, regardless of whether their suggestions are adopted or not.

Cities are responsible for 75% of the world's climate emissions (UN, 2023). Therefore, actively working to reduce urban emissions is crucial, whether through changing technologies, altering habits, or introducing new technologies that can cut down emissions.

Conclusion

This chapter has explored the multifaceted concept of smart cities, scrutinizing the critical role citizens play in their development. Throughout our analysis, we have encountered various interpretations of what constitutes a smart city, from technological advancements to social and environmental initiatives. Central to our discussion is the acknowledgment that the definition and purpose of smart city solutions significantly influence both the social consequences and the level of citizen involvement in their implementation.

In smart city planning and development, we observed that the interests of various stakeholders often diverge. While providers of smart city technology aim for widespread adoption of their solutions, and academia seeks projects that attract research funding, residents predominantly focus on solutions enhancing their personal well-being. These dynamics underscore the importance of balancing the different agendas to ensure smart city solutions align with the diverse needs of the community.

The Quadruple Helix Model emerged as a particularly relevant framework, emphasizing collaboration among government, industry, academia, and civil society. However, this model is not without its challenges. It struggles with power imbalances and the complex task of integrating diverse stakeholder perspectives, which can lead to conflicts and decision-making inefficiencies. Our analysis revealed that the extent of citizen involvement varies significantly across different smart city projects, influenced by the city's political system, the technology's objectives, and the balance of power among stakeholders.

The question of when and why citizen involvement may be limited or necessary in implementing smart city technology also arose. This led to a discussion on the merits of liberalistic versus paternalistic governance approaches in the context of smart cities. We considered examples like Halden, Norway, where budget constraints and the complexity of smart city solutions pose challenges to citizen involvement, and scenarios where early citizen involvement could mitigate resistance to new technologies.

Ultimately, our exploration highlighted the indispensable role of citizen involvement in the successful development of smart cities. Whether through public consultations, participatory design, co-creation workshops, or pilot projects, involving citizens ensures that smart city technologies align with their real-life needs and expectations. This approach not only enhances the effectiveness of smart city solutions but also fosters a sense of community ownership and engagement in urban development.

As we move forward, it is essential to recognize that smart city development is not a one-size-fits-all approach. The diverse needs of citizens, the unique characteristics of each city, and the ever-evolving nature of technology demand a flexible, adaptive, and inclusive approach to urban planning. The role of citizens, therefore, is not just as beneficiaries of smart city initiatives but as active participants shaping the future of urban living. In this light, the journey toward developing smart cities is as much about technological innovation as it is about fostering democratic engagement, ensuring equity, and enhancing the quality of life for all residents.

A fundamental principle to follow is that smart city technology should be employed to benefit the residents. Consequently, residents should have a say in deciding not only which technologies are implemented but also in how they are applied.

References

Arnkil, R., Järvensivu, A., Koski, P., & Piirainen, T. (2010). *Exploring quadruple helix – outlining user-oriented innovation models*. Finland: University of Tampere, Institute for Social Research, Work Research Centre.

Arnstein, S. R. (1969). A ladder of citizen participation. *Journal of the American Institute of Planners*, *35*(4), 216–224.

Batty, M., Axhausen, K. W., Giannotti, F., Pozdnoukhov, A., Bazzani, A., Wachowicz, M., Ouzounis, G., & Portugali, Y. (2012). Smart cities of the future. *The European Physical Journal Special Topics*, *214*, 481–518.

Boon, L. S., Malek, J. A., Hussain, M., & Tahir, Z. (2017). *Citizen participation in realising the citizen-centric vision for smart city*. Social, environmental and developmental sustainability research centre international conference.

Carayannis, E. G., & Campbell, D. F. (2009). 'Mode 3'and 'Quadruple Helix': Toward a 21st century fractal innovation ecosystem. *International Journal of Technology Management*, *46*(3–4), 201–234.

Carayannis, E. G., & Campbell, D. F. (2010). Triple Helix, Quadruple Helix and Quintuple Helix and how do knowledge, innovation and the environment relate to each other?: A proposed framework for a trans-disciplinary analysis of sustainable development and social ecology. *International Journal of Social Ecology and Sustainable Development (IJSESD)*, *1*(1), 41–69.

Carayannis, E. G., Campbell, D. F., Carayannis, E. G., & Campbell, D. F. (2012). *Mode 3 knowledge production in quadruple helix innovation systems: Twenty-first-century democracy, innovation, and entrepreneurship for development*. Springer.

Carayannis, E. G., & Rakhmatullin, R. (2014). The quadruple/quintuple innovation helixes and smart specialisation strategies for sustainable and inclusive growth in Europe and beyond. *Journal of the Knowledge Economy*, *5*, 212–239.

Commission, E. (2023, November 10). *Smart cities*. https://commission.europa.eu/eu-regional-and-urban-development/topics/cities-and-urban-development/city-initiatives/smart-cities_en#:~:text=A%20smart%20city%20is%20a%20place%20where%20traditional,technologies%20for%20better%20resource%20use%20and%20less%20emissions.

Dahl, R. A. (2008). *Democracy and its critics*. Yale University Press.

De Jong, M., Joss, S., Schraven, D., Zhan, C., & Weijnen, M. (2015). Sustainable–smart–resilient–low carbon–eco–knowledge cities; making sense of a multitude of concepts promoting sustainable urbanization. *Journal of Cleaner Production*, *109*, 25–38.

Fung, A. (2006). Varieties of participation in complex governance. *Public Administration Review*, *66*, 66–75.

Gracias, J. S., Parnell, G. S., Specking, E., Pohl, E. A., & Buchanan, R. (2023). Smart cities—A structured literature review. *Smart Cities*, *6*(4), 1719–1743. https://www.mdpi.com/2624-6511/6/4/80

Griffy-Brown, C., Earp, B. D., & Rosas, O. (2018). Technology and the good society. *Technology in Society*, *52*(1), 1–5.

Habermas, J. (1985). *The theory of communicative action: Volume 1: Reason and the rationalization of society* (Vol. 1). Beacon Press.

Helbing, D., Fanitabasi, F., Giannotti, F., Hänggli, R., Hausladen, C. I., van den Hoven, J., Mahajan, S., Pedreschi, D., & Pournaras, E. (2021). Ethics of smart cities:

Towards value-sensitive design and co-evolving city life. *Sustainability (Basel, Switzerland)*, *13*(20), 11162. https://doi.org/10.3390/su132011162

Helsinki, C. O. (2024). *The Helsinki of dreams is created together*. OmaStadi. https://omastadi.hel.fi/?locale=en

Hollands, R. G. (2014). Critical interventions into the corporate smart city. *Cambridge Journal of Regions, Economy and Society*, *8*(1), 61–77. https://doi.org/10.1093/cjres/rsu011

Hollands, R. G. (2020). Will the real smart city please stand up?: Intelligent, progressive or entrepreneurial? In *The Routledge companion to smart cities* (pp. 179–199). Routledge.

IESE. (2024). *IESE cities in motion index 2024*. B. S. o. Navara. https://www.iese.edu/media/research/pdfs/ST-0649-E

IMD. (2024). *IMD smart city index 2024*. I.-I. I. f. M. Developm. https://issuu.com/docs/e7a60c053affbf9e98fcba93afe857af?fr=xKAE9_zU1NQ

Innes, J. E., & Booher, D. E. (2004). Reframing public participation: Strategies for the 21st century. *Planning Theory & Practice*, *5*(4), 419–436.

Ivanova, I. A., & Leydesdorff, L. (2014). Rotational symmetry and the transformation of innovation systems in a Triple Helix of university–industry–government relations. *Technological Forecasting and Social Change*, *86*, 143–156.

Jaeger, P. T. (2011). *Disability and the Internet: Confronting a digital divide*. Lynne Rienner Publishers.

Kitchin, R. (2022). Conceptualising smart cities. *Urban Research & Practice*, *15*(1), 155–159. https://doi.org/10.1080/17535069.2022.2031143

Komninos, N. (2014). *The age of intelligent cities: Smart environments and innovation-for-all strategies*. Routledge.

KS. (2023). *Legally regulated citizen participation*. https://www.ks.no/fagomrader/demokrati-og-styring/innbyggermedvirkning/om-innbyggermedvirkning/lovregulert-innbyggermedvirkning

Leydesdorff, L., & Etzkowitz, H. (2001). The transformation of university-industry-government relations. *Electronic Journal of Sociology*, *5*(4), 1–17.

Mason, A. (2020). *Nordic-digital-municipality-index-2020*. https://www.telenor.com/binaries/about/public-policy/Nordic-Digital-Municipality-Index-2020.pdf

Mutambik, I. (2023). The global whitewashing of smart cities: Citizens' perspectives. *Sustainability*, *15*(10), 8100. https://www.mdpi.com/2071-1050/15/10/8100

Nam, T., & Pardo, T. A. (2011a). *Conceptualizing smart city with dimensions of technology, people, and institutions*. Proceedings of the *12th annual international digital government research conference: Digital government innovation in challenging times*.

Nam, T., & Pardo, T. A. (2011b). *Smart city as urban innovation: Focusing on management, policy, and context*. Proceedings of the *5th international conference on theory and practice of electronic governance*.

Nordrum, E. I. (2023). The technologically sustained digital divide. In *Technology and sustainable development* (pp. 97–108). Routledge.

Pile, S. (1999). What is a city. In *City Worlds*, edited by John Allen, Doreen Massey, and Steve Pile (3–52). London: Routledge.

Porter, M. E., & Heppelmann, J. E. (2014). How smart, connected products are transforming competition. *Harvard Business Review*, *92*(11), 64–88.

Romanelli, W. C. M. (2019). *Power relationships in the co-production of smart city initiatives*. ItAIS 2019 - The XVI conference of the Italian Chapter of AIS, Naples (Italy). https://www.researchgate.net/publication/336208993_Power_relationships_in_the_co-production_of_smart_city_initiatives

Sætra, H. S. (2019). Freedom under the gaze of Big Brother: Preparing the grounds for a liberal defence of privacy in the era of Big Data. *Technology in Society*, *58*, 101160.

Sætra, H. S. (2020). A shallow defence of a technocracy of artificial intelligence: Examining the political harms of algorithmic governance in the domain of government. *Technology in Society*, *62*, 101283. https://doi.org/10.1016/j.techsoc.2020.101283

Sanders, E. B.-N., & Stappers, P. J. (2008). Co-creation and the new landscapes of design. *Co-Design*, *4*(1), 5–18.

Smartcities.org. (2022). *Smartcities index report*. https://smartcitiesindex.org/service innovation

Syed, A. S., Sierra-Sosa, D., Kumar, A., & Elmaghraby, A. (2021). IoT in smart cities: A survey of technologies, practices and challenges. *Smart Cities*, *4*(2), 429–475. https://doi.org/10.3390/smartcities4020024

UN. (2015). *Sustainable development goals*. United Nations Department of Economic and Social Affairs. https://sdgs.un.org/goals

UN. (2023). *Cities and climate change*. Environment Programme. https://www.unep.org/explore-topics/resource-efficiency/what-we-do/cities/cities-and-climate-change

Vanolo, A. (2016). Is there anybody out there? The place and role of citizens in tomorrow's smart cities. *Futures*, *82*, 26–36.

Wadhwa, T. (2015). Smart cities: Toward the surveillance society? In *Smart cities as democratic ecologies* (pp. 125–141).

2

INCLUDING DISADVANTAGED CITIZENS IN SMART SOCIETIES

How learning in formal and informal educational settings fosters digital literacy

Alexander Smit, Maud Rebergen, Joëlle Swart, and Marcel Broersma

Introduction

In an era where digital literacy is a fundamental prerequisite for participating in society, understanding how low-literate and low digitally literate citizens navigate digital spaces is critical for fostering inclusive smart societies. In practice, these groups are nevertheless often excluded as more and more societal processes become digital-by-default and ultimately smart (Berry and Glaeser, 2005; Friemel et al., 2021; Ince, 2022). While the Netherlands is considered one of the front runners in terms of digitization, approximately one in five Dutch citizens aged between 16 and 74 is lacking essential literacy skills (Buisman et al., 2023). Moreover, almost one in four Dutch citizens is lacking basic digital literacies (Rijksoverheid, 2023). Despite the increased need for digital literacies in the smart society, there is limited knowledge about how digital literacies are developed, in particular how these are understood and applied in formal and informal educational settings. Moreover, the role of digital capital in relation to the development and application of digital literacies in practice is understudied (Ragnedda et al., 2024).

Hence, this chapter explores how formal and informal educational settings contribute to or impede the development and application of digital literacies among disadvantaged public in smart societies, analyzing two case studies through a Bourdieusian (1986) lens: (1) an informal digital literacy initiative in a community center primarily leveraging social capital and (2) a formal digital education program in a vocational school primarily leveraging cultural capital. Employing participant observations and semi-structured interviews with learners, volunteers, and educators, we examine the situated socio-material contexts influencing digital literacy development and analyze

DOI: 10.4324/9781003439325-4

how both spaces differently affect the acquisition of digital capital (see Ragnedda, 2018). We do so by taking a human-centered, emic approach towards the concept of smart cities, highlighting citizens' own experiences regarding digital technologies, digital education, and digital inclusion to inform smart city initiatives (Zaidan et al., 2022).

Digital literacy has become key to participate in both public and private spheres, including domains like electronic government (e-government) and electronic commerce (e-commerce) (Adam & Alhassan, 2021; Ince, 2022). The advent of smart technologies has also given rise to a digital participatory culture, requiring a more diverse set of skills that relate to broader cultural, civic, and economic competencies, such as expression, creation, sharing, interaction, and engagement (Meyers et al., 2013; Aabo, 2005). As Tham and Verhulsdonck (2023) argue, stimulating such skills for all citizens is fundamental to the success of smart city initiatives. This success should be gauged by the inclusiveness and usability of smart technologies, which require the provision of accessible digital education, both formal and informal, to nurture an informed and digital literate citizenry.

However, despite their importance for participation in the smart city, a significant portion of the population remains devoid of basic digital literacy skills (Helsper, 2021; Van Dijk, 2020). This hinders people's ability to utilize many technological resources, meaning they run the risk of being left behind. Furthermore, digital literacy pedagogies often only relate to formal educational contexts, neglecting the significance of informal spaces for developing, appropriating and enacting digital literacies in everyday life (Smit, Swart & Broersma, 2023). Hence, frictions arise between formalized digital literacy initiatives and situated informal digital literacy practices.

This chapter therefore compares the development and application of digital literacies within both types of educational environments. Employing two case studies, we analyze how each contributes to the formation of inclusive smart societies. Understanding the different notions of digital literacy development in formal vis-à-vis informal educational spaces enables us to design tailored educational resources that nurture digital literacies and digital capital in context-specific settings.

The chapter starts by situating digital literacies in relation to smart technologies and digital capital. We discuss how formal and informal educational settings affect the enactment of digital literacies by introducing two case studies that draw on different educational frames. First, we present our findings from our participant observations and interviews at a local community center, exploring how informal educational environments embed more affective, situated pedagogies that make them better suited for disadvantaged public. Second, we explore digital literacy development at a school for vocational education, demonstrating how formal educational environments embed more technical and material learning outcomes. Finally, differences in both pedagogical perspectives are analyzed, evaluating their positive and negative attributes in terms of digital literacy development among disadvantaged public. In the conclusion, the added value of these insights is discussed highlighting that digital literacy, in relation to digital capital, is a fundamental factor for effective participation in smart societies. We argue that it transcends basic technical skills, acting as a crucial enabler of citizen engagement with digital innovation.

Situating digital literacies

In a technology-driven world, characterized by the proliferation of smart cities and the Internet of Things, digital literacy have emerged as foundational condition (Castells, 2011; Gilster, 1997; Ince, 2022). Current smart technologies require citizens to be able to not only access, but also leverage digital technologies in different modalities, spaces, and contexts, and understand their material properties and functionalities with diverse interrelated sets of skills and competences (Ince, 2022; Tham & Verhulsdonck, 2023). In this sense, digital literacy can thus be conceptualized as the extent to which individuals are able to harness diverse digital technologies to competently seek, assess, generate, and disseminate diverse forms of information from an array of digital sources (Tinmaz et al., 2022). While often perceived as a competency primarily nurtured within formal educational settings, digital literacy is also cultivated in a variety of informal learning environments like the home, streets, and similar community spaces (Meyers et al., 2013; Potter, 2020; Smit, Swart & Broersma, 2024).

Moreover, the responsibility for developing digital literacy, previously viewed as predominantly the domain of educational institutions, is now recognized as a collective obligation of all learning environments, both formal and informal (Potter, 2020). Diverse educational contexts are vital in equipping individuals of all ages with digital literacy skills and facilitating their continuous development and renewal (Meyers et al., 2013; Potter, 2020). As Information and Communication Technologies (ICTs) and smart technologies grow in complexity, they demand more varied and interconnected digital literacies, especially prevalent in both informal and formal settings.

Scholars have pointed out that digital technologies may perpetuate current power imbalances and reproduce social inequalities in the digital realm (Goedhart, 2021; Ragnedda et al., 2020; Van Dijk, 2020). According to Saraceni (2020), this unequal distribution of social power attributes to an uneven proliferation of skills and competences needed to gain benefits from smart city technologies. Issues of access, skills, and limited digital participation arise when individuals face barriers on the basis of these issues (Helsper, 2021; Van Deursen & Van Dijk, 2019). Hence, we understand digital literacy as a conditional aspect of digital capital.

Digital capital is conceptualized as a distinct form of capital that can be accrued and transformed into various other forms of more 'traditional' capital in the social world (see Bourdieu, 1986; Ragnedda, 2018), thereby enhancing individuals' life opportunities. Digital capital comprises 'internalized', or learned socio-digital competencies in relation to 'external', or material features of digital technologies (Ragnedda, 2018). We argue that digital capital is unique in its capacity to act as a conduit between online and offline spaces, which is especially relevant in smart societies as online and offline spheres are merged. Moreover, possessing digital capital is essential for effectively leveraging other forms of capital – such as economic, cultural, human, and social – in the digital domain (Ragnedda et al., 2020), and to foster potential benefits and increase life-chances and inclusion in contemporary smart cities for diverse public. Hence, it should be understood as 'the dominant form of capital in the digital field' (Ragnedda, 2018, p. 56).

Digital capital is intrinsically linked to the concept of the 'digital divide', with the absence of the former contributing to the persistence of the latter. Current research on digital inequalities highlights three distinct digital divides, resulting in different levels of access and frequency of technology use; different types of access, user skills, and diversity of usage; and different outcomes and benefits of media use (Helsper, 2021; van Dijk, 2020). The first divide concerns economic-based access disparities, often seen as a divide between those who have and those who do not, but also the quality and depth of access (Hargittai, 1999; Goedhart, 2021). The second divide deals with differences in digital skills and usage types, categorized as motivational, material, skills, and usage access (van Dijk, 2020). The third divide focuses on the varied results people achieve from using ICTs (Helsper, 2021). All these divides may be addressed through educational initiatives, but as Hilbert (2011) points out, the perception of what constitutes a digital divide is not universally agreed upon among policymakers. Consequently, he proposes that, rather than a uniform approach, situation-specific strategies may be more effective in addressing both the digital divide and digital literacy programs. Such situation-specific strategies can be leveraged to embed formal and informal educational structures into a holistic and more fundamental approach toward digital inclusion.

Formal and informal learning spaces

Contemporary technological innovation necessitates a reevaluation of the distinction between formal and informal learning contexts (Dezuanni et al., 2023). The pervasive nature of digital technologies, facilitating a continuous state of connectivity regardless of physical location, underscores the need to view each learning process as simultaneously formal and informal (Alper, 2017). However, traditionally, studies on digital literacy education have been mostly focused on formal learning spaces (Bawden, 2008). This perspective fails to encapsulate the dynamic and everyday cultural aspects of digital life. It overlooks the complex, mobile, and layered contemporary digital space and its interrelation with multiple offline environments, which are more effectively integrated in informal environments like the home or public community spaces.

One of the main differences between formal and informal educational environments traces back to traditional literacy education (Potter, 2020). Within the 'new literacy studies' framework (Street, 2003), two primary literacy education systems are identified: the autonomous model, which focuses on mastering technical and functional abilities, and the ideological model, which offers a socially situated understanding of literacy. The latter recognizes literacy as a process influenced by socio-cultural norms and communicative practices, aligning more closely with informal contexts. The contemporary landscape of digital literacy, characterized by its dynamic and multimodal nature, necessitates an educational approach that adapts to shifts in meaning-making, intertextuality, and the interaction of different modes. This challenge involves creating a curriculum that embraces the complexity and fluidity of digital literacy in the modern age. Informal learning environments are crucial in applying and developing such adaptive digital literacies, contributing to forms of 'life-wide' learning (Meyers et al., 2013) beyond the more rigid and motivationally challenging structures of formal education.

An integrated perspective to fostering the digital literacies necessary to participate in smart cities recognizes both informal and formal settings as part of a comprehensive knowledge ecosystem. In everyday life, people engage with digital technologies across various activities and spaces, illustrating the seamless nature of learning and literacy in the digital world. The distinction between formal and informal educational settings in digital literacy is not about elevating one over the other. Rather, it aims to highlight the importance of physical and digital spaces that are often overlooked in academic discourse, but are essential for how digital literacy, digital capital, and digital inclusion are developed (Potter, 2020).

This chapter draws upon participant observations and semi-structured interviews conducted between 2019 and 2021 in one formal and one informal learning space to analyze how digital literacies develop, considering affect, social support structures, different sorts of capital, and barriers for developing digital literacies (see Spradley, 2016). Employing this qualitative mixed-method approach enabled us to study from a user-centric perspective how digital literacies are learned, practiced, and enacted in practice. Integrating observations with interviews allowed us to not only observe digital practices in real-time but also elicit participants' reflections on these practices afterward. This approach helped overcome some of the inherent limitations of relying solely on interviews as we could directly observe digital actions, moving beyond the reliance on retrospective accounts and enhancing the depth of our understanding.

As an informal context, we focus on a local community center in the Oosterpark neighborhood in the city of Groningen, the Netherlands, where we observed and interviewed three volunteers and eight students working on advancing digital skills. Key aspects investigated within the informal educational environment included how low-literate people were reached and engaged in digital literacy education, how pedagogies are situated in social relations, the affective dimension of digital education, and the relationships between volunteers and learners.

Our case study for a formal learning space is a school for vocational education in Groningen, Alfa College, where we observed and interviewed 23 native Dutch (NT1) and 29 non-native (NT2) Dutch citizens following a digital skills course. This case study centered on students' experiences of more structured curricula, examining how such formal frameworks influenced digital literacy development and how they compared to more informal developments. The observations focused on how low-literate adults understood and practiced digital literacies within informal and formal educational settings. The observations followed the participant observation model put forth by Spradley (2016), moving from more general to focused observations.

The interviews focused on learners' comprehension of digital media and the development of digital literacies in relation to everyday life. For example, more formalized instances of digital literacy, preferences for specific devices and applications (e.g., smartphones with WhatsApp), frequency of use, motivations for acquiring digital literacies, and how educational contexts embedded these preferences were discussed. This aimed to uncover underlying user motivations and affective dispositions. Interviews were conducted in-person and typically lasted around 40 minutes, concentrating on participants' use of media despite language barriers. We also discussed how the volunteers as educators understood digital literacy, what their experiences and vision on digital pedagogies were, how they thought about working with disadvantaged low-literate students, and what their positive and negative experiences with digital media education entailed. In the semi-structured interviews with the students, the focus was on how they thought about digital literacy education, how it related to their situated and affective needs, barriers, and experiences with digital media, and how they valued these informal educational spaces.

All observations were documented by means of field notes in a research journal. The interviews were audio-recorded and transcribed, after which all data was coded in Atlas.ti using thematic analysis (Braun & Clarke, 2012). This involved an iterative coding process where initial codes were created, followed by another round of coding for all transcripts to develop more detailed codes reflecting the recorded cultural settings. This process was repeated four times to construct a comprehensive and valuable representation of the data. Pseudonyms are used to reflect the data in an anonymous manner.

Informed consent was obtained from all participants in either written or oral form. Occasionally, acquiring written informed consent was challenging due to participants' difficulties with reading and writing, particularly for those who did not speak Dutch or English. This was addressed by carefully obtaining oral consent. The researcher introduced themselves, explained their role and the research topic, sometimes using translation software, to secure informed consent in written or oral form. Participants who could not give their informed consent were excluded from the study. The study received final approval from the Ethics Board of the University of Groningen (CETO) on May 17, 2023, under reference number 572–93593093.

Informal educational setting: Community Center

This community center is located in a neighborhood in the northeastern part of Groningen, with a population of around 10,000, and is characterized by a higher proportion of elderly, young people, and non-Western immigrants compared to the rest of the city (CBS, 2023). The area generally exhibits lower levels of welfare and education. Despite recent investments, the district still faces challenges like unemployment, poverty, and precarity (CBS, 2023). However, it also demonstrates a relatively strong social cohesion and is active in initiating projects to mitigate these issues.

In this community center, social gatherings that contribute to the social cohesion of the neighborhood are hosted. A notable element of this informal support structure is the role of volunteers. These local residents have an intrinsic motivation for volunteering in the digital skills educational program. They indicate that they like to expand their social network within the community, fulfill personal time commitments, or contribute meaningfully to their neighbors' well-being. Many volunteers chose to teach at the language center (in Dutch: taalhuis), an informal space to learn different literacy practices, seeking personal connections with less digitally skilled groups from Oosterpark, as described by Anna (F, 51, interview volunteer 2): '"I first worked at a vocational school, but saw that the students weren't always motivated and didn't want to be there. Here at the community center, everybody is so happy and motivated to learn something, I really love that'. This commitment from volunteers plays a crucial role in reinforcing and preserving informal networks and social structures within the community. This voluntariness also accounts for the visitors of the community center, as they are free to come by or leave as they want, in contrast with formal education where students are mostly obliged to be present.

Reaching and engaging disadvantaged low-literate public

Successfully reaching disadvantaged individuals presents a challenge for many institutions, social agencies, and researchers (SLS, 2022). A key factor in the success of the Oosterpark community center is the minimal physical and social distance between them and the target group. Volunteers residing within the neighborhood possess a situated understanding of the local challenges and share an affinity with their neighbors, thereby lowering barriers to participation and enhancing motivation for both parties in educational programs, as described by Harma (F, 57, interview volunteer 1):

> If I ask a disadvantaged community member to go to a school, that person almost naturally says: no that isn't something for me because of the formal setting. However, if I invite that person to our community center for a cup of coffee, that is fine and that person does not feel a large barrier towards coming over because of the informal social setting.

Volunteers engage with participants in informal settings, like having coffee or taking walks in the neighborhood. This direct method of engagement is particularly crucial for reaching low-literate individuals, a group that is typically hard to connect with through traditional online or offline advertising methods. The dual role of the community center as both a social hub and a gateway to (informal) education is vital in accessing such disadvantaged citizens in need of assistance. The close connection volunteers have with participants also motivates them to help, as they witness firsthand the impact of low literacy and limited digital skills in their community.

The community center serves as a pivotal gateway for both low-literate and digitally low-literate individuals to engage informally in language and digital skills education. This accessible and welcoming environment proves highly effective in reaching disadvantaged groups, such as low-literate individuals, facilitating their introduction to both general and digital education. Conversely, it is also an effective way to recruit potential volunteers, as Harma (F, 57, interview volunteer 1) in the community center says:

> I want to be useful to my community members, and I like to do that in the neighborhood. We also live in the Oosterpark neighborhood. I like to help my peers a little further with knowledge that I myself have at hand. Moreover, it also expands your network a little in the neighborhood. So, well, let's say it is a win-win for everyone.

This approach generates a cascading effect where neighbors inform one another about educational initiatives, facilitating organic growth, social capital, and outreach within the community. Volunteers extend their roles beyond the confines of the classroom, often acting as a personal supporting party, feeding into important support structures for inclusion and participation (see Smit, Swart, and Broersma, 2024).

Situated affective pedagogies

The value of informal education largely stems from its ability to tailor programs for engaging with the personal contexts of participants. Volunteers have also noted that the informal nature of the educational setting enhances participant motivation and being heavily reliant on this relationship built on trust and equality, which is vital for engaging and sustaining the interest of the target group. Safety within this relationship allows for meaningful discussions about media usage, past unpleasant experiences, and tactics for dealing with these issues in daily life. Emma (F, 53, interview volunteer 3) mentions:

> I often have one or two community members that come by when I'm at work, because they know me and know that I don't judge their lack of literacy or digital literacy. We are equal and that creates a safe environment.

Disadvantaged groups, such as those with low literacy levels, feel more heard and empowered in these settings, where individual attention is prioritized, and learning objectives are pursued in smaller groups or even individually. In the development of digital capital, the significance of social capital and resources is paramount (Ragnedda, 2018). Hence, a strong social foundation, characterized by trust, familiarity, confidence, and safety, is crucial, particularly for marginalized groups. Without this social underpinning, the cultivation of digital capital becomes more precarious and less anchored. This is shown by Maria (F, 49, interview volunteer 4):

> When people come in that are referred by their friends, they are often motivated and trust you as a volunteer to help them. If someone comes in without being referred, we notice that the emotional barrier of trusting us is way higher.

Hence, social resources in relation to trusted social actors and affective barriers are essential in how educational spaces are experienced. Additionally, tailored education, responsive to the specific and personal needs of learners, coupled with an understanding of their everyday (digital) challenges, emerges as highly effective. Such an approach enables educators to align with the participant's current level and discern their needs for digital advancement.

Individuals with low literacy levels place significant importance on the specific type of medium utilized in digital spaces. Their digital participation depends on their prior experiences and levels of confidence, trust, and familiarity in relation to the materiality of hardware and software. In their everyday activities, they commonly use Android smartphones, due to their accessibility, relative cost-efficiency, and intuitive interface. However, the digital skills acquired in this context do not seamlessly transfer to using other digital devices, like laptops or smartphones with other operating systems and increased complexity. This indicates a need for tailored educational approaches that consider these variances in media familiarity and media anxiety (Ince, 2022). For example, one participant indicated that their learning process was greatly disrupted due to the lack of an external computer mouse being attached to a laptop and the participant being forced to use the internal touchpad of the laptop:

> You could notice that if I had had a mouse I would have done the assignment far more quickly. That makes a huge difference for me. I had to use that laptop without a mouse and with that thing [touchpad]. I'm not used to that thing, it only makes it more difficult and I get frustrated by it. I had a lot of trouble with it and it didn't go as I'm used to. At home this assignment is fairly easy, but here with these different things it's far more difficult.
>
> *(Mike: M, 56, interview student 12)*

Engaging with unfamiliar media or devices and their technical properties poses significant challenges in participation, persistence, and motivation for digital activities, particularly when users experience a loss of control due to the difficulty of operating technological features. For example, Mohammed (M, 47, interview student 9) explained that he is scared to use a computer, as he does not understand all its features and is afraid of doing something wrong. At the same time, he likes using his smartphone, as he knows its primary functions and trusts himself more with the device. In the realm of digital literacy development, habitual use of digital devices fosters confidence among digitally low-literate individuals, who typically struggle with uncertainty, unfamiliarity, and lack of trust in their digital skills. Regular interaction with media enables them to develop intuitive, affective, and sensory understanding of hardware and software, as evidenced by Sara (F, 41, interview student 7): 'I love using my smartphone as it gives me a lot of options to use it in a way that I don't have to read a lot of text or have this technical understanding'. Educational programs can harness this foundational knowledge to cultivate digital literacies and, consequently, enhance digital capital.

The role of volunteers

One drawback to informal education is that volunteers are often not experts in their fields and are mostly not pedagogically trained to educate others.

This poses limits on the levels of professionalism of digital literacy education and support, as Anna (F, 57, interview volunteer 2) explains: 'I would like to have someone that I can contact to learn about Word or Excel [...] those are very complicated programs and I only know the very basics'. Volunteers also suffer from limited time, funds, and sometimes may have other priorities. It is also problematic that such structures are very dependent upon these volunteers, and the support structure crumbles when volunteers are unavailable, or when volunteers have different priorities. Therefore, they can often only educate others at a very basic level in ad-hoc manners. In the interviews, volunteers indicated that they did not know if they could answer complex questions, should they arise. Several volunteers indicated that they would need more information, support, and training on digital literacies to efficiently help participants. This would make them feel better prepared for their tasks and be able to further develop themselves in terms of their digital skills. For example, the following volunteer described how she thinks about potentially having to use software like Excel:

> I'm not really familiar with Excel or any of those things, so if someone asks a question about it I'm not sure what to say. I once took a course in that, but I hardly ever use it. (...) If I have to know all the ins and outs of that, I might have to go on a course first, but the possibilities to do that are very limited.
>
> *(Harma: F, 57, interview volunteer 1)*

Moreover, volunteers often use their own digital devices to educate others, which in some cases include outdated hardware and operating systems. This situation presents risks to the educational process, such as the dissemination of obsolete information and potential instability and security issues with the hardware and software. This issue is particularly prevalent in digital literacy education as digital technologies and the skills to employ them effectively are subject to fast-paced innovation. In some cases, volunteers did not have the right software licenses and felt they had nowhere to turn with their ICT-related questions:

> No, I don't have a laptop anymore, no. And neither does the municipality have one for me [...]. They used to rent such devices for volunteers like me, but don't structurally do such things anymore. But I have had problems before and then I said: somebody has to come and help me with this. They often react with 'no, we do not update or account for such devices, you have to figure it out yourself'.
>
> *(Maria: F, 39, interview volunteer 3)*

In examining the impact of volunteers' limited capacity on the effectiveness of informal digital literacy education, it becomes evident that these constraints can lead to less effective learning outcomes and create access barriers, limiting skill development opportunities. This highlights the importance of strengthening the foundational aspects of informal education, particularly in terms of access. A critical aspect of this discussion is the digital capital of volunteers, encompassing both access and skills. While there may be deficiencies in this regard, volunteers leverage social capital and resources to create a learning environment characterized by solidarity, and a human-centric approach. This contrasts more formal educational settings, where cultural capital is prioritized.

Formal educational setting: Vocational school for digital skills building

The second case study focuses on a school for vocational education in Groningen, where participants followed a digital skills course integrated into a language course for low-literate adults, both people having Dutch as their native language (NT1) and people not possessing Dutch as their native language (NT2), e.g., immigrants. The combination with language teaching is fruitful here, because applying digital skills and dealing with digital technology often requires language skills as well. An adapted teaching program for digital low-literates had been developed on the basis of five domains on digital skills drawn up by the Ministry of Education, Culture and Science: (1) using ICT systems, (2) security, privacy and health, (3) searching for information, (4) processing information, and (5) digital communication. The lesson series had different starting levels, depending on the students' level of digital literacies. On the initiative of the teachers, the lessons paid relatively much attention to the theme of 'digital security'. In addition, students had the opportunity to propose their own topics or learning objectives. At the end of the 20-week program, participants could take an exam in which the learning objectives of various domains are tested.

Becoming digitally literate

What participants indicated they wanted to learn during the digital literacy course was very diverse and strongly related to their individual contexts and life circumstances. Examples included learning how to write a job application letter in a word processor with the correct formatting, how to fill out and send online forms from public agencies, how to do online banking, and how to install and use a computer or a printer. Parents also expressed motivations to gain control over their children's digital activities. For example, one student indicated that she would like to be able to monitor her children on the computer, to gain more insight into their (digital) world and keep them safe:

> Online gaming is dangerous. (...) If you want to go to a higher [level], others might say do this or that. Or he is [in] contact with scary people that I don't know. What's his or her name? I don't know. My son and my daughter sometimes learn bad words. I was surprised. Sometimes, for example, my son plays and sometimes bad words are said. In the house we do not use such words, so when I ask: What is this, how do you know these words? He answers that he learned it online (...) I want them to have privacy. (...) But sometimes I have to make sure they are safe online.
>
> *(Ashra, F, 36, interview student 3)*

This shows that digital literacy is not only related to individual circumstances but is also important in relation to monitoring and surveilling digital acts of children, especially as parents are responsible for their children in offline and online settings, a responsibility not accompanying other social actors in their social circle. It also shows a desire to control something that, in their perspective, is unknown and distant. Other participants expressed the feeling that they felt the need to become digitally literate in order not to be excluded and to continue to participate in a society that is increasingly focused on digital technology:

> So every month, for example, I have to forward a copy of my paycheck to social services. If I don't do this, I can lose my allowance. And downloading, uploading and scanning documents and all those things, that's just a problem for me with possibly large consequences.
>
> *(Asaf: M, 43, interview student 5)*

For some participants, digitization appears as an abstract concept that introduces additional barriers rather than resolving existing ones. In these cases, the absence of digital literacy often results in a lack of cultural, social, and economic capital, exemplified by the potential loss of allowances due to the inability to submit digital documents. Consequently, this exacerbates social stratification, widens the digital divide, and intensifies digital inequality. Digital literacies are thus perceived as essential external requirements for functioning in smart societies, rather than voluntary personal skills.

Educators emphasize the need to clearly communicate the benefits of digital literacy courses to learners, particularly those with limited digital knowledge and experience. The value of digital literacies is not always evident to the digitally illiterate, creating a paradox where individuals need digital literacy to understand its significance in acquiring various forms of capital. Digital literacy education should therefore aim to align with learners' practical needs and goals, combining general learning outcomes with personal objectives and integrating learners' daily digital experiences. This necessitates teachers to be flexible and have a wide range of digital skills, along with the ability to adapt formal learning outcomes to more informal, everyday contexts. Continuous professional development for educators is essential to keep them abreast of new technologies and applications, thus enabling them to provide tailored and personalized support.

Material familiarity

As mentioned above, for disadvantaged groups, smartphones often serve as the primary, and often only, access point to the digital world due to their relative affordability and intuitiveness. As a result, people develop digital literacies closely tied to the features and affordances of smartphones, such as touchscreens and simplistic interfaces. Many, for instance, do not own computers at home due to economic constraints, yet are required to learn digital skills on PCs in educational settings. While this may offer a broader understanding of digital media, the skills acquired often do not translate directly to their personal, smartphone-centric digital experiences. However, formal digital literacy courses typically involve using computers, introducing a dissonance between learned skills and real-life applicability for these participants. This is in line with our findings, where most participants were found to manage reasonably well on smartphones but struggled with PCs or laptops. They perceived smartphones as easier and more intuitive, underscoring a disparity in their level of familiarity and proficiency across different digital devices:

> Yes I can manage that, I understand that with my phone. If I want to send an email via my phone, I get it. I don't know how to do it on my laptop, that's way more difficult.
>
> *(Hans: M, 46, interview student 5)*

This highlights the comfortable relationship with personal devices and the challenges faced when using unfamiliar technology, such as school PCs. The fear of making mistakes and their consequences, and of the unknown, often leads to avoidance of learning resources and diminished confidence. Despite available teacher support, many students hesitate to ask for help due to this apprehension, preferring assistance from their social networks as they know their peers are equal and understand their situated barriers. This reflects a gap in accessible support within the learning environment and a distrust in unfamiliar external resources.

Given the necessity for proficiency with various digital devices, leveraging dynamic digital literacies is essential. Learners' existing smartphone skills provide a valuable starting point for educators. By building on this existing knowledge, instructors can introduce more complex digital literacy concepts, illustrating their relevance across different technological platforms.

This method capitalizes on learners' current abilities, progressively broadening their digital literacy scope.

Barriers to developing digital literacy

Our findings point to two main barriers to developing digital literacy in formal educational environments: anxiety and shame. Most of the students indicated at the start of the course that they lacked self-confidence and were afraid of doing something wrong if they were to practice using the computer independently. That is why they felt anxious at the start of the course, sometimes even with palpitations as a result. Although this anxiety diminished during the course by gaining experience with digital devices, the lessons did not completely eliminate these emotions. This led, among other things, to the avoidance of some applications of digital technologies. For example, one student indicated that she did not want to learn online banking because she found it too risky:

> I'm afraid of that. (...) Suppose I do it wrong, then I think I will not understand the consequences, I don't know. That the money is gone then too and I will have big problems.
>
> *(Rosa: F, 47, interview student 11)*

The student described that she would rather not participate in such digital activities entirely, or ask a familiar and trusted peer, like a family member or friend, to outsource this task.

Second, feelings of shame emerged as a significant barrier for participants, who expressed embarrassment about their limited computer skills at the start of the course. Although this uncertainty decreased over time, it persisted throughout the curriculum. The participants' lack of computer experience highlighted their reliance on social support, underscoring the importance of social capital and resources (Bourdieu, 1986; Helsper, 2021). Establishing a support network is crucial for developing digital literacy (Friemel et al., 2021). However, some individuals faced the challenge of not having friends or family to assist with digital tasks, compounded by their reluctance to seek help due to shame or discomfort. Educational courses can counteract these issues by fostering peer connections and mutual support. By doing so, these courses help participants recognize that they are not alone in their digital challenges, thereby facilitating a shared understanding of these difficulties and promoting collective learning.

Furthermore, these formal environments can inadvertently engender a sense of detachment between educators and learners, adversely affecting the learning experience, due to enacting digital media like computers or laptops and learning digital literacies that are not transferable to informal contexts like the home when such devices are not present there. This issue is especially acute for learners who necessitate additional motivation to initiate and sustain their educational engagement. The findings suggest that a more heterogeneous and flexible educational approach, which deviates from rigid curricula to concentrate on participants' personal experiences, is more conducive to fostering digital literacy development. This underscores the importance of adapting educational strategies to suit the varied needs and backgrounds of learners, particularly in settings where building digital competencies is a key objective.

Differences in developing digital literacies

Figure 2.1 visualizes the primary differences between informal and formal educational environments on the basis of how digital media education in our case studies is experienced by the low-literate participants and shows the characteristics and differences of both contexts in relation to the development of digital literacies. The key distinction lies in how informal and formal educational environments approach digital literacy. Informally, digital literacies are deeply intertwined with social practices and rely on social support structures, trust, equality, and positive affirmation, aligning with Street's ideological model of literacy education (2003). These literacies are mostly learned and internalized through affective dispositions and social resources.

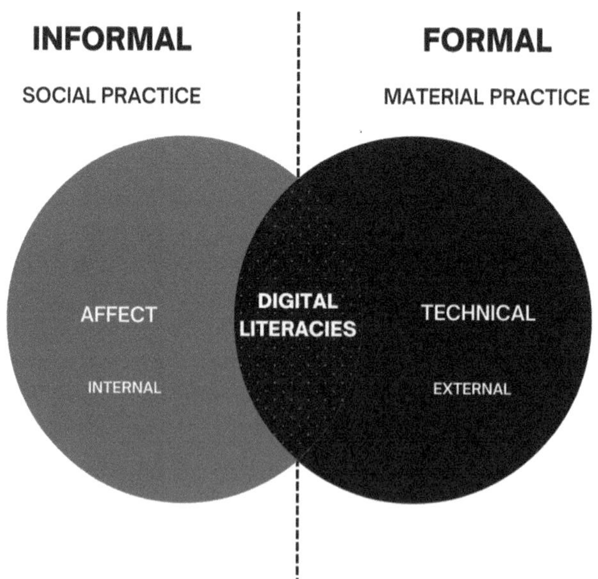

FIGURE 2.1 Informal and formal educational environments.

Conversely, formal educational settings connect digital literacies more with material practices and attribute them to the technical aspects of ICTs, reflecting Street's autonomous model of literacy (2003). Here, literacy is embedded in an instrumental, techno-solutionist framework, and feelings of fear and shame are prominent barriers. This environment emphasizes the external dimension of digital capital, focusing on 'external' material properties of ICTs in contrast to the 'internalized' socio-digital competencies of informal spaces (see Figure 2.1). Formal education excels in delivering comprehensive, structured instruction on the technical and material aspects of digital tools, fostering a deeper understanding of digital literacies and smart technologies. Such pedagogies are crucial for equipping individuals with the digital literacies necessary for engagement in smart city contexts.

The presence of trained educators facilitates the delivery of more complex curricula, offering structured and well-planned programs that provide a sense of security and support. Furthermore, formal education often culminates in an official certificate, serving as tangible evidence of the acquired digital skills. However, formal education in our case study falls short in addressing the individual needs and life circumstances of those with low digital literacy, and hence can be qualified as linear, standardized, and homogenous. Its emphasis on generic skills and the broad scope of its programs limit the capacity to cater to the unique personal contexts of digitally low-literate individuals. This gap highlights the need for a more tailored approach within formal educational frameworks to better meet the diverse requirements of this group.

Informal education proves highly effective for developing basic digital literacies, particularly for individuals with limited digital skills. Its success lies in its adaptability to each learner's unique circumstances, abilities, and objectives. This approach facilitates a trust-based relationship between volunteers and participants, thereby enriching the learning experience. Volunteers view informal education as a platform that fosters community connections through a primarily social approach. The role of social capital and resources is foundational, not only in addressing practical needs within smart cities but also in cultivating positive affirmation, trust, safety, and familiarity. However, volunteers also acknowledge the superiority of formal education in advancing digital literacies at a more comprehensive level, attributed to the presence of trained educators and structured curricula and pedagogies. Volunteers often depend on their immediate networks for digital assistance, but these networks may lack adequate digital capital.

Therefore, it is essential to enhance volunteer support and provide additional training to ensure the effectiveness of informal education. Uncertainties regarding their digital competencies can adversely affect the quality of instruction, impacting both volunteer motivation and participant learning outcomes. Addressing these challenges requires bolstering support for volunteers, including the provision of appropriate hardware, software, and continuous training. In addition, it fosters a learning space with a low threshold, accessible to all sorts of community members to learn and develop different digital skills in a safe, equal, trusted, and motivated manner, often integrated in a social activity like drinking a cup of coffee and talking about everyday topics. This approach is crucial in optimizing their educational roles and ensuring the effective delivery of digital literacy education in informal settings.

By aligning closely with participants' daily experiences and challenges, informal education offers an efficient and relevant means to address digital difficulties in everyday life. Key to its effectiveness are the consistent structure, repetition, and use of familiar devices and media, especially crucial for low-literate individuals. These strategies provide insights into the digital challenges faced by vulnerable groups and serve as a foundation for volunteers to foster educational and social connections. The insights from informal education can greatly enhance formal digital literacy programs.

Despite its strengths, informal education faces challenges like limited resource access and outdated technology. The choice of media and its alignment with the learners' socio-cultural, economic, and personal contexts are essential components of the educational experience. Therefore, maintaining consistency in the type of media used throughout the educational process is important. This consistency reduces the need for learners to continually adjust to new technologies, thus optimizing the learning process.

Conclusion

Our findings highlight that digital literacies, in relation to digital capital, are a fundamental factor for effective participation in smart societies. They transcend basic facilitative skills, acting as a crucial prerequisite of citizen engagement in an increasingly digitalized society. However, the research also shows a disparity in the recognition and value attributed to informal learning environments in contrast with formal education. This could negatively impact the development and dissemination of digital literacies and, consequently, digital inclusion and participation in everyday life. The underlying paradox entails that informal structures are more in line with everyday digital life in smart cities and represent a lower threshold for participatory action. However, most of these informal structures are not transparent to institutions, pedagogies, teachers, and more. For example, when they are acted upon in the home setting behind closed doors, they are often assumed to be non-existent. At the same time, these more socially situated digital literacies seem to be the primary point of departure for low-literate people to participate in smart societies (see Smit, Swart, and Broersma, 2024). In practice, however, these skills are limited and often tied to a specific digital medium, like the smartphone, consequently impacting their options for digital participation. Hence, people ultimately need a more substantial and dynamic scope of digital literacies acquired in formal and informal settings to be able to digitally participate in smart cities and navigate more complex digital infrastructures and platforms. The question arises how we can solve this challenge.

Our study adds valuable perspectives on this complex interplay of inclusionary and exclusionary mechanisms in the development and enactment of digital literacies. It emphasizes that informal education, when customized to the needs of participants, notably enhances digital literacy, particularly in contexts of social relevance. We show that the success of informal education can be attributed to its personalized, practical focus and its local, accessible, and community-based setting. This advances our understanding of developing digital literacies in informal educational spaces that, as argued by Meyers et al. (2013), serve as a critical link between the digital and the social and thus enable further skill and knowledge advancement in situated manners (Mensonides et al., 2024). These findings highlight the importance of a holistic approach that synergistically combines the advantages of both formal and informal educational methods to maximize learning efficacy.

Another key insight is the importance of addressing intrinsic motivation in the target group, centering their interests, needs, desires, and personal reasons for developing digital skills, and leveraging their media familiarity and prior experience to learn additional digital literacies. Educators can harness these motivations to show the relevance and opportunities of digital media, linking it to individual interests and the affective aspect of digital literacy development. This would also propel the advancement of digital capital, as argued by Ragnedda (2018), when both internal and external dimensions of media usage and understanding would strengthen one another and foster the acquisition of digital capital, enabling digital participation and ultimately facilitating digital inclusion.

This understanding advances the argument of Ragnedda (2018), as our findings show that digital access is not simply a question of whether a skill or technology is present or absent, but is about a citizen's capacity for engagement in public life within a specific situated, affective, and social configuration, rooted in digital capital. Drawing from this understanding, we advocate for a nuanced understanding of digital literacies as a relational condition for digital participation, emphasizing the role of context, technology usage, and societal values. It calls for a synergistic approach that combines formal and informal learning, focusing on promoting digital literacy and inclusion, especially among disadvantaged subgroups. Furthermore, a deeper exploration of digital literacy is required, moving beyond oversimplified models to consider the complexities of digital media consumption, production, and education. This involves looking at the broader social context beyond the individual and their immediate environment. This perspective acknowledges the fluid boundaries of literacy and learning and urges scholars, educators, and institutional collaborators to appreciate the multifaceted and interlinked nature of literacy practices. Furthermore, it scrutinizes the structural factors that foster an unequal distribution of digital literacies among the citizenry with a diverse subset of functional literacies. This dynamic relational aspect should be central to our understanding of digital literacies and digital capital in the smart society.

Acknowledgments

This work was supported by the Dutch Research Council (NWO), grant nr. 410.19.008 (https://www.nwo.nl/en/projects/41019008). We would also like to thank all the participants for sharing their experiences, perspectives and opinions with us. We would also like to thank the libraries, community center and the educational institution for their help with recruiting participants for this study and allowing us to gather data at their facilities.

Funding

The author(s) also disclosed receipt of the following financial support for the research, authorship, and/or publication of this article: This publication is part of the research project 'Informed Citizenship for All. Digital Literacy as Prerequisite for an Inclusive Society', funded by the Dutch Research Council (NWO), grant no. 410.19.008, and supported by the National Library of the Netherlands (KB), the Ministry of the Interior and Kingdom Relations (BZK) and Stichting Kinderopvang Stad Groningen (SKSG).

References

Aabø, S. (2005). The role and value of public libraries in the age of digital technologies. *Journal of Librarianship and Information Science*, *37*(4), 205–211.

Adam, I. O., & Alhassan, M. D. (2021). The linkages between ICT access, e-government and government effectiveness and its effect on corruption. *International Journal of Technology, Policy and Management*, *21*(4), 344–362.

Alper, M. (2017). *Giving voice: Mobile communication, disability, and inequality*. MIT Press.

Bawden, D. (2008). Origins and concepts of digital literacy. *Digital Literacies: Concepts, Policies and Practices*, *30*(2008), 17–32.

Berry, C. R., & Glaeser, E. L. (2005). The divergence of human capital levels across cities. *Papers in Regional Science*, *84*(3), 407–444.

Bourdieu, P. (1986). The forms of capital. *Cultural Theory: An Anthology*, *1*(81–93), 949.

Braun, V., & Clarke, V. (2012). *Thematic analysis*. American Psychological Association.

Buisman, M., Bollen, I., Jacobs, B., Huijts, T., Cornelisse, R., & van Guilik, N. (2024). PIAAC 2023.

Castells, M. (2011). *The power of identity*. John Wiley & Sons.

Centraal Bureau voor de Statistiek (2023). *Demografische gegevens op lokaal niveau.* Statline.

Dezuanni, M., Osman, K., Foth, M., Mitchell, P., McCosker, A., Notley, T.,... & Mavoa, J. (2023). Digital inclusion is everybody's business: Key findings from the ARC Linkage Project Advancing digital inclusion in low-income Australian families.

Friemel, T. N., Frey, T., & Seifert, A. (2021). Multidimensional digital inequalities: Theoretical framework, empirical investigation, and policy implications of digital inequalities among older adults. *Weizenbaum Journal of the Digital Society*, *1*(1), 1–24.

Gilster, P. (1997). *Digital literacy*. New York: Wiley Computer Pub.

Goedhart, N. S. (2021). *Social inclusion in digitizing societies: Starting from the lifeworld of people with a low socioeconomic position.* PhD-Thesis - Research and graduation internal, Vrije Universiteit Amsterdam. ProefschriftMaken. (see https://research.vu.nl/en/publications/social-inclusion-in-digitizing-societies-starting-from-the-lifewo)

Hargittai, E. (1999). Weaving the Western Web: Explaining differences in Internet connectivity among OECD countries. *Telecommunications Policy*, *23*(10–11), 701–718.

Helsper, E. (2021). The digital disconnect: The social causes and consequences of digital inequalities. *The Digital Disconnect*, 1–232.

Hilbert, M. (2011). Digital gender divide or technologically empowered women in developing countries? A typical case of lies, damned lies, and statistics. *Women's Studies International Forum*, *34*(6), 479–489.

Ince, F. (2022). Digital literacy training: Opportunities and challenges. In M. Taher (Ed.), *Handbook of research on the role of libraries, archives, and museums in achieving civic engagement and social justice in smart cities* (pp. 185–199). IGI Global. https://doi.org/10.4018/978-1-7998-8363-0.ch009

Mensonides, D., Smit, A., Talsma, I., Swart, J., & Broersma, M. (2024). Digital literacies as socially situated pedagogical processes: Genealogically understanding media, information, and digital literacies. *Media and Communication*, *12*, 8174.

Meyers, E. M., Erickson, I., & Small, R. V. (2013). Digital literacy and informal learning environments: An introduction. *Learning, Media and Technology*, *38*(4), 355–367.

Potter, J. (2020). Digital literacy/'Dynamic Literacies': Formal and informal learning now and in the emergent future. In *The Routledge companion to digital media and children* (pp. 256–264). Routledge.

Ragnedda, M. (2018). Conceptualizing digital capital. *Telematics and Informatics*, *35*(8), 2366–2375.

Ragnedda, M., Ruiu, M. L., & Addeo, F. (2020). Measuring digital capital: An empirical investigation. *New Media & Society*, *22*(5), 793–816.

Ragnedda, M., Ruiu, M. L., & Calderón-Gómez, D. (2024). Examining the interplay of sociodemographic and sociotechnical factors on users' perceived digital skills. *Media and Communication*, *12*.

Rijksoverheid (2023). *Geactualiseerde Werkagenda Waardegedreven Digitaliseren.* Ministerie van Binnenlandse Zaken en Koninkrijksrelaties.

Saraceni, G. (2020). Digital divide and fundamental rights. *Human(ities) and Rights Global Network Journal*, *2*(1), 66–91. https://doi.org/10.24861/2675-1038.v2i1.27

Smit, A., Swart, J., & Broersma, M. (2023). Digital inclusion of low-literate adults: Challenging the sequential underpinnings of the digital divide. In *Weizenbaum conference "Practicing sovereignty: Interventions for open digital futures"* (pp. 72–84).
Smit, A., Swart, J., & Broersma, M. (2024). Bypassing digital literacy: Marginalized citizens' tactics for participation and inclusion in digital societies. *New Media & Society*, 0(0), 1–19. https://doi.org/10.1177/14614448231220383
Spradley, J. P. (2016). *Participant observation*. Waveland Press.
Street, B. (2003). What's "new" in new literacy studies? Critical approaches to literacy in theory and practice. *Current Issues in Comparative Education*, *5*(2), 77–91.
Tham, J. C., & Verhulsdonck, G. (2023). Smart education in smart cities: Layered implications for networked and ubiquitous learning. *IEEE Transactions on Technology and Society*, *4*(1), 87–95.
Tinmaz, H., Lee, Y. T., Fanea-Ivanovici, M., & Baber, H. (2022). A systematic review on digital literacy. *Smart Learning Environments*, *9*(1), 1–18.
Van Deursen, A. J., & Van Dijk, J. A. (2019). The first-level digital divide shifts from inequalities in physical access to inequalities in material access. *New Media & Society*, *21*(2), 354–375.
Van Dijk, J. (2020). *The digital divide*. John Wiley & Sons.
Zaidan, E., Ghofrani, A., Abulibdeh, A., & Jafari, M. (2022). Accelerating the change to smart societies-a strategic knowledge-based framework for smart energy transition of urban communities. *Frontiers in Energy Research*, *10*, 852092.

3

FROM A SMART CITY TO WISE CITIZENS – SMART EMPOWERING WITH ARTIFICIAL INTELLIGENCE

Kristina Khutsishvili

Introduction

The concept of smart cities has been a pillar for reflection on urban technological development and the consequent urban future studies context (Etezadzadeh, 2016). Yet, as with every conceptual foundation, it has also been approached critically, with a common argument of a one-sided, too techno-centric view, bringing into the light the means of 'smartness' instead of citizens (here and after meaning the inhabitants of cities rather than the nationals) and their communities. This may also be seen as part of the debate on technology and society: while seeking the balance, one may see reciprocity being the feature of societal-technological coexistence, when technological developments respond to the demands of society and influence the processes in the society itself with the consequent societal change. In times of rapid technological developments, such as nowadays with the focus on artificial intelligence (AI) and particularly the large language models, the view on reciprocal relation between technology and society may swift into the primary and even deterministic role of technology (Hughes, 1994).

On par with the human- and community-centric arguments, the exploratory work has been done aimed at other directions and priorities, such as resilience, with resilient cities being seen as a necessity in the context of global challenges (see, for example, the Resilient Cities Network established in 2013; Open and Agile Smart Cities broadening the focus as Open and Agile Cities and Communities, 2023). The community focus oftentimes relates to the Sustainable Development Goals and the recent Covid-pandemic experiences. 'How do we build sustainable communities? Yes, not just sustainable

DOI: 10.4324/9781003439325-5

cities, but communities'" ask the leaders of the International Federation of Consulting Engineers in the relevant report (2021, 2).

The essential character of communities has come to the forefront of the city-centred initiatives. For example, the smart cities' and communities' vector is outlined by the European Commission as a vector of work aimed at both tackling the local challenges and reaching the European Green Deal objectives (European Commission's Digital Strategy, Smart Cities and Communities, 2022). The community focus is relevant to the city-centred technological initiatives, such as CitCom Artificial Intelligence (AI), co-funded by the European Commission Testing and Experimentation Facility (TEF) aimed to test AI and robotics solutions for smart communities. Unlike smart city, the concept of smart community is not thoroughly elaborated. The European Commission defines both smart cities and communities through the shared aim of 'well-being of its inhabitants, businesses, visitors, organisations and administrators by offering digitally enabled services that contribute to a better quality of life' (Shaping Europe's Digital Future: Smart Cities and Communities, 2022).

The similar trend may be noted in the AI community, with the importance of human-centricity articulated and practical solutions being tried out. Humanness, citizenship, the exercise of agency, and participation are nowadays mentioned in discussions in both knowledge domains and on the intersection of those – the AI and, more broadly, new emerging technologies aimed to respond to the needs of cities and communities. This intersection, on the one hand, brings together the members of the world of technology with citizens and communities, and on the other, it comprises what belongs to the domains of social and technological innovation. Bringing together the theoretical reflection and the practical findings deriving from the CommuniCity Horizon Europe project, in this chapter I argue for the reciprocity in the translation and co-creation effort in the technological domain and for the mutual empowerment of technological providers and citizens and communities. In this way, the knowledge transfer goes 'both ways' and may be seen as mutual learning, and the social and technological innovations for the cities and communities are viewed not in silos but intertwined.

Participation

The empowerment function of AI is possible in the conditions when the preliminary work is done, meaning that the citizens should be familiarised with and accustomed to the digital 'toolbox'. The digital literacy level is a precondition for benefiting from the rapid development of technological solutions. The facilitative function hence may be seen as the responsibility of the city's policymakers, while the knowledge-sharing function can be assigned to universities, research centres, but also the research and development departments of technological companies that now oftentimes lead the exploratory efforts on the fast-paced innovation frontiers. The factor of moral responsibility, involving knowledge sharing, should not be undermined. Continuity in this sense is necessary: with the speed of the development of AI-based technologies, not only the involved professionals need to keep up with the pulse of change but also citizens, as the digital literacy aim is now not a 'fixed' point of competence and being at 'ease' with technology but rather the context for an open-ended knowledge and experience sharing process.

The necessary conditions of participation are the translation efforts, in a metaphorical but oftentimes also literal sense. The community-centred view on the city and the citizens may be seen as having philosophical underpinnings, traced in works of communitarian scholars such as Charles Taylor (1989, 1992). In the context of immigration and globalisation, community translation is a necessity in the context of tackling the digital divide and making smart empowerment of citizens possible. The proposed definition of community translation, 'a service offered at a national or local level to ensure that the members of multilingual societies have access to information and active participation' (Taibi & Ozolins, 2016, 8), zooms the specificity. While the act of translation may be seen as universal, the specific – place-based, local – component of community translation needs to be recognised. It may be seen in both literal and metaphorical sense: each community having its specific needs, varying levels of awareness on technology-mediated opportunities and challenges, different levels of trust in technology, its providers, and facilitating organisations. The 'specific' should not be undermined by the global, universal trend of technological development.

In the context of technological knowledge and know-how, one of the parties engaged in the translational efforts is the party having the understanding of the concepts, tools, framework of innovation, and the relevant knowledge needed for another party to be equipped for the participatory processes. In an attempt to formalise, such a disposition may be seen as a two-system structure with an unequal power relation: the asymmetry of power constituted in technological knowledge favours the provider of technological knowledge.

This situation may be seen as 'fair', as the knowledge acquiring is aimed to benefit the citizens and their communities, to prepare them for productive and mutually rewarding participation. The parallel here can be made with the pattern of knowledge acquisition at educational institutions, with unequal power relations being the norm rather than an exception, and with the goal to prepare the students for future participation in society and at their chosen workplaces. Yet, the social innovation focus should not be forgotten.

The view on the city- and citizen-centred initiatives, with a special focus on innovation, benefits from the holistic standpoint taken, meaning the inclusion of both technological and social innovation vectors, seeing them not in silos but coexisting, interrelated, intertwined. This adds value to shaping the realistic picture of the processes in cities and communities and more nuanced and better-informed discussion. With the social and technological innovation domains being viewed inside one common framework, the issue of translation starts to be seen differently. The power structure is more balanced: with the attention to the vector of social innovation, the knowledge of the needs of citizens and communities comes to the forefront, and hence the structure gets more symmetry, with the knowledge coming from citizens and communities including their experiences and needs being valued and the efforts of translation going 'both ways'. This not only addresses the power imbalance but also facilitates the conditions for mutual recognition, with the latter having identitarian consequences (Taylor, 1992).

Reciprocity

The reciprocal relation may be viewed as the necessity for fair AI aimed at the goals for social good. Not only the AI systems have the potential to empower citizens, but also the citizens have the ability to tackle the issues related to AI systems, with the issues of bias being the primary example. AI enables and empowers the citizens and enhances their capabilities; the latter may be seen inside the human enhancement paradigm (Khutsishvili, 2024b), and the citizens in turn add value in making AI more fair. Co-creation of AI solutions with citizens may be seen as means for tackling the problem of bias and developing more fair by design technological solutions.

Co-creation is not an easy task: it needs to be supplemented by extensive preliminary work, the assumptions may be challenged by the real-life processes, and the questions of reward and motivations are not answered unequivocally. Co-creation in the technological domain requires preliminary efforts, including the efforts of (metaphorical but sometimes also literal) translation of the concepts, tools, principles – in sum, the knowledge that needs to be shared in order to bring the parties closer to each other. Not only the technological knowledge and know-how are meant at this point but also the understanding of the civic, citizen-centred perspective by the technological provider or any other 'techno-centric' party. The one-sided approach of technological knowledge sharing is not sufficient: it perpetuates the conditions for power asymmetry in the co-creation space, as if only the technological provider has the knowledge that other parties including the co-creators may not have. In practice, the technological provider also oftentimes lacks knowledge – for example, the knowledge of the needs of citizens and communities. The assumption that such knowledge is not significant or is less significant for the success of the process would perpetuate the situation of power imbalance in the technological domain.

The best and simply good enough practices of co-creation of technological solutions with members of target communities may be presented in different forms, be it formalised tools understandable for more tech-inclined audiences and narrative guidelines, with references to empirical data deriving from experiments and its analysis (Khutsishvili, 2024a). Such practices would not only go in line with the declared principles of 'tech for social good' but may also become both formalised and narrated tools aimed at knowledge sharing, replication, and interoperability in the European Union.

The concepts of fairness and justice exceed the scope of the studies on AI and machine learning, with multiple angles that can be relevant: philosophical, computer science-driven, social science-driven, etc. The emotional component is very much present in the fairness discussions (a question: what is 'fair', not only on the level of definition, but in its essence – what is 'fair' for one agent may or may not be viewed as 'fair' by another). In application to AI, the interdisciplinarity challenge is present as well: translation from one discipline to another, the need for mutual understanding, the socio-technical nature of the phenomenon.

Possible definitions of fair AI also vary: we can see it as a system treating people in a just way/impartially/in a non-discriminatory/non-biased way; as a decision support system that avoids disparate benefit or harm to different groups; as a framework that 'builds upon a predefined sensitive attribute (e.g., race, gender, disability) that describes membership in a protected group' that 'should not lead to discrimination' (Feuerriegel et al., 2020, 381); as the idea that 'any two individuals who are similar with respect to a particular task should be classified similarly' (Dwork et al., 2012, 1).

In the scope of literature on co-creation, the specificity of co-creation in the technological domain has not been elaborated extensively, with attempts suggesting the need for a different approach to the issue itself when it comes to co-creating technological solutions with citizens (Jarke, 2021, 203; Spuzic et al., 2016). Even more so, the existing literature lacks the focus on engagements with members of vulnerable and marginalised communities, where ethical aspects and risk mitigation have even more significance.

Co-creation processes are approached by a wide range of academic disciplines, including management and innovation studies that aim to give a structured overview of the issue rather than providing a conceptual framework. For instance, Roser et al. (2013, 23) define co-creation as 'an interactive, creative and social process between stakeholders'. Within the context of innovation studies, co-creation is seen as 'open innovation with customers' (Rayna & Striukova, 2015, 38). It 'corresponds to the customer-related part of open innovation: "open innovating" with consumers necessarily implies co-creating with them' (Rayna et al., 2015, 2).

Co-creation of technological solutions with members of target communities is an example of the bottom-up design approach, with the data generated by the community simplifying the efforts towards algorithmic fairness and tackling the problem of bias. On the other hand, community involvement, especially in cases of disadvantaged groups, implies a stress on the 'do no harm' principle and a careful ethical attitude. There are no universal 'receipts' on how to facilitate a productive and all-involved-parties rewarding collaboration with no risks of disappointment and/or harm. While the goal to establish best practices of co-creation with citizens in technological domain, be it a narrative form of the guidelines or an attempt of the formalised toolkits, is reasonable and worth the efforts to make, the ethical consideration in every individual case is an immanent part of the process.

Practice and lessons learned

On the European level, we may point out different practical attempts of 'smart empowering' of citizens by means of AI and broader new emerging technologies. One of the informative tools of knowledge sharing in this domain is the Living-in.EU movement – the platform launched in 2019 by the Open and Agile Smart Cities and Communities (OASC), Eurocities, the European Network of Living Labs (ENOLL).

CommuniCity is one of the European projects aimed at smart empowering by means of AI and broader new emerging technologies. From the practical perspective, CommuniCity had been designed as a European framework, with a timeframe of three years starting in September 2022, to facilitate and fund 100 technological pilots – involving AI and other new emerging technologies – in European cities, with the core cities being Amsterdam, Helsinki, and Porto. The first piloting round targeted these three cities, while the second round was also expanded to the so-called 'replicator' cities – Prague, Aarhus, Tallinn, and Breda. The focus of the project is on vulnerable and marginalised communities and their needs to which the technological solutions are aimed to respond. The participating cities are supervising the awarded pilots in which the technological providers are involved as well as the associations – intermediaries helping to get in touch with the target communities.

Among the lessons learned during the first piloting round of CommuniCity project in Amsterdam in relation to smart empowering with AI, we can outline:

- For a technological provider, it is important to thoroughly explore the needs of the target community as a part of preparatory work; establishing the contact with target group may be not easy; getting to know people also takes time;
- Co-creation methods can be used not only to develop a technological solution but also to connect with the target community;

- A ready solution cannot always be adjusted to the needs of each particular community: here comes the question of the 'universal' and the 'specific'. While we do focus on coming up with the replicable practices, it is important to always keep an eye on peculiarities of the context;
- The more vulnerable the target community is, the higher the risk that harm can be caused in case when the expectations are not met;
- When a technological provider starts a pilot in collaboration with an association (e.g., a non-profit organisation), it is essential to confirm that the association has the established connections with the target community;
- In some cases, it may be wise to start co-creation activities right away if the technological provider feels that the connection with some members of the target community is made, rather than to wait until a bigger group of participants is established; while waiting for others, the technological provider may lose the contact that is already made;
- It is important to ensure that the co-creators are available from the beginning of the pilot; otherwise the piloting timeframe may not be enough to finalise activities;
- 'If I have the impression I am not being listened to, I drop out'; 'We have to look for a working solution together. The company should not do it alone. I want to be taken seriously' (target community members); it is important to recognise the effort made by community members; mutual recognition is a key for successful co-creation;
- If the technological provider 'tests' the solution with the members of target community, it is important to have a follow up when the participants are informed how their feedback was used, what changes were made; in this scenario, the participants; they can see the impact of their input and check whether their feedback was understood correctly by the technological provider;
- According to some participants, being a 'test person' can increase self-confidence;
- It is important to listen well, take participants seriously, avoid stereotyping or taking superior position; appreciation of the time investment is important; participants should not feel exploited; respect, recognition, and equality in communication are important; information should be shared and follow up provided: participants should know how their input or feedback was used;
- The pilot host (e.g., the municipality) should not exclude an option of stopping the pilot if things are clearly not going as expected and the problems are serious enough;
- The tips from the civil servants: be yourself, be vulnerable, build a connection; make participants feel comfortable; be 'equal' to your target community, not 'above' the participants; don't hesitate to moderate the discussion

and ensure that the quiet participants get a chance to speak their minds too; build trust;

- Don't underestimate the necessary effort on behalf of the technological provider (a remark from a provider: 'Technology and animation are particularly difficult, and we need to explain a lot. However, no matter how extensively we explain, more and more questions are coming');
- Something to think about: if it is decided to remunerate the participants, how to ensure that people participate not only because of the remuneration?
- It is important to be in touch with the needs of the end users, not only in the beginning of the pilot but also during the whole process;
- Some projects may need more time and funding to produce sustainable results and involve the target community better;
- Project-culture versus problem-culture: most of the times the project cycle is too short to address all issues for a problem to be solved;
- More moments in which 'the whole system is in the room', including the members of target community, to prevent miscommunications and relying on associations too much; in case if the target community consists of vulnerable people, time needs to be invested to prepare the participants;
- For a technological provider it is important to test the solution thoroughly, validating whether everything works and is useful; implementation of functionalities that do not work well yet may foster disappointment, frustration, distrust in the target community so intended users may lose interest and curiosity;
- Expectation management is important when working with the target community: participants may expect the solution to work out straightaway and may lose interest if the reality does not match the expectations; thorough and honest communication and explanation is necessary;
- Technological providers need to be explicit and clear and make sure that both the associations they are collaborating with and the members of target communities understand that technological solutions they are working on are in the process of development and not yet finished. For people who are not used to the development of technology, this is not obvious; participants and collaborators may expect things to work straightaway. Technological provider needs to explain that they need help of people from the target community to make their product or project work;
- The situation when technological providers are not physically located in the proximity of target communities may bring some challenges, as offline co-creation/co-design/testing meetings with target communities are important;
- The 'translation' from the articulated challenge to the technological solution needs attention: it is important to keep an eye that the solution that is being developed meets the needs of the target community;

- In a particular pilot, a person from the target community felt that the transition from the first conversation to being 'used' as a translator in the literal sense was too fast and abrupt; the efforts should be made to prevent the members of target communities feel exploited; collaboration with communities should be mutually rewarding;
- The power dynamic between municipalities and members of target communities should also get attention;
- In a particular pilot, a participant from the target community was worried that technology may not be able to deal well with emotions related to the process in which the tool is aimed to be used; he was not sure whether developers are aware of this issue. The focus on technological solution may 'blind' the developers from considering the risks pointed out by members of community;
- In a particular pilot, a person from the target community felt the developers do not realise well enough that he is just one Arabic speaker, a particular person, not a 'general audience'. He emphasised that he belongs to a particular class, is highly educated and in general cannot represent the whole target community. He is not sure that the developers are fully aware that their target community is not a monolith;
- The issue of ownership is important: communities do not appreciate the municipalities 'taking over' too often when community members have a good idea;
- There may be a problem of trust towards the municipality in target communities: 'I first have to sell myself, then the programme, then the municipality' (municipality worker).

The accumulation of learnings aimed at replication of the solutions in other cities and communities is the main goal of CommuniCity project. One of the main learnings applied to the second and third rounds of the open calls is bringing the needs of community members at the centre of the process design: without sincere interest and in-depth immersion in people's life circumstances, methods, approaches, and know-how will likely remain formal tools (Khutsishvili et al., 2024).

Conclusion

Smart empowering with AI and broader new emerging technologies implies the necessity of knowledge accumulation and consequent participation. The translation and co-creation efforts are essential features of the context with reciprocal relations in place: the context in which the processes of empowering citizens and communities by means of technology coexist on par with participation of citizens and communities in the development of new technologies, making those less biased and more fair 'by design'.

The reciprocity has not only an economic or governance-based justification but also a meaningful value-based underpinning: not only the citizens can be empowered by means of technology, but also the technological providers can be empowered by getting closer to the citizens and communities and their needs. With this understanding, the need-based approach has been at the core of the CommuniCity project. Finding out and analysing the needs of communities is a part of work preceding each new round of open calls; the needs then are transferred into the challenges articulated in the call announcements, to which the applicants are aimed to respond.

The need-based approach does not diminish the importance of the grant-awarded applicants to connect with the target communities themselves. Moreover, the lack of contact is recognised as one of the factors that may determine the problems during piloting and the lack of success of a given pilot. When the sincere interest is seen at the end of a technological provider or any other involved organisation, communities usually are keen to respond, with their input being needed both for the responsible and fair technology development and for the well-functioning of facilitating organisations such as municipalities, the innovation companies of the cities, and other stakeholders with similar goals.

Multiple opportunities and challenges brought by rapid technological developments require deliberation, which, in turn, is based on participation. It would be interesting to review and re-assess the corpus of literature on democratic participation, so the ideas discussed in that context can be 'translated' and explored in the context of challenges and opportunities brought by new emerging technologies, including the pressing issues of governance and regulation. It is not only important to ensure that the social and technological innovations are viewed not in silos but intertwined, but to link the discussions on democratic participation and co-creation, so the already accumulated knowledge can be valuable.

Co-creation practices may be viewed as part of the general democratisation and participation trend, and, as it is with other democratic practices, not only the voice matters but all activities and processes preceding the voice being raised. Fruitful participation needs preparation aimed at the necessary knowledge and 'toolbox' acquisition that enables and facilitates participation. Preparation efforts should be seen as essential part of co-creation processes: performance at this stage has a potential to influence or even determine the outcome of the whole endeavour. The lack of attention may disregard any significant effort put further into the co-creation sessions itself. The negative outcome then can range from economic implications, as co-creation sessions imply the costs, to the disappointment of participants and decrease of trust in target communities.

In case co-creation is encouraged by the governmental institutions, such as municipalities or innovation companies of the cities, as it is in CommuniCity project, the issues of trust are crucial. The well-founded element of a problem is the lack of trust in governmental institutions within vulnerable communities; it is already there and is aimed to be tackled by the encouraged participation. The scenario when improper facilitation of participation exacerbates the serious problem it was aimed to tackle should be avoided.

The members of target communities have their initial expectations from participation; they put effort that should be recognised. The format of such a recognition has been debated in the CommuniCity project, with some participants articulating the 'must' of financial remuneration of the efforts of co-creators, and with others making the argument of 'not right' motivation taking place when remuneration is the driving force of participation. This is an example of a practical discussion in which it is not easy to conclude one of the parties being right and another being wrong.

Co-creation is not easy; it requires a lot of effort from the parties involved and a thorough preparation needed to address the risks of things going wrong. For that, knowledge transfer and efforts of translation are necessary. While in the technological domain the specific knowledge leads to the asymmetry of power, with a more powerful position of a technological provider, the knowledge of communities and their needs is no less important. The translation efforts hence should be directed not only from technological providers to community members but also from community members to technological providers. The latter also includes the understanding of the needs of this specific community to which a provider is aimed to respond. Such a reciprocal recognition and mutual learning should be seen as necessary features of successful empowerment action in the technological domain in cities and communities.

Acknowledgements

The lessons learned in Amsterdam are a result of intense and highly collaborative work, with the city of Amsterdam being the center and the driving force of it. I would like to acknowledge the highest level of dedication, creativity, and integrity demonstrated by Neeltje Pavicic and Machteld Combé, the city of Amsterdam, in the CommuniCity project, that made our productive collaboration a pleasure.

This work has been supported by the European Union's Horizon Europe research and innovation programme under grant agreement no 101070325, project CommuniCity Innovative Solutions Responding to the Needs of Cities & Communities.

References

CitCom AI: Testing AI in smart cities and communities, project under the Digital Europe programme. https://citcom.ai; https://digital-strategy.ec.europa.eu/en/activities/testing-and-experimentation-facilities

CommuniCity Horizon Europe Project. https://communicity-project.eu
Dwork, C., Hardt, M., Pitassi, T., Reingold, O., & Zemel, R. (2012). Fairness through awareness. *Proceedings of the third innovations in theoretical computer science conference* (pp. 214–226). https://www.cs.toronto.edu/~toni/Papers/awareness.pdf
Etezadzadeh, C. (2016). *Smart city 2.0 as a livable city and future market*. Springer Vieweg.
Eurocities. https://eurocities.eu
European network of living labs (ENOLL). https://enoll.org
Feuerriegel, S., Dolata, M., & Schwabe, G. (2020). Fair AI. *Business and Information Systems Engineering*, *62*(4), 379–384. https://link.springer.com/article/10.1007/s12599-020-00650-3
Hughes, T. P. (1994). Technological momentum. In M. R. Smith, & L. Marx (Eds.), *Does technology drive history? The dilemma of technological determinism*. MIT Press.
Jarke, J. (2021). Co-creating digital public services for an aging society: Evidence for user-centric design. In *Public administration and information technology* (p. 6). Springer.
Khutsishvili, K. (2024a). Guidelines for translating frameworks, methods, tools and principles of local innovations for marginalised and vulnerable communities – 2023. *Open Research Europe*. https://open-research-europe.ec.europa.eu/articles/4-36
Khutsishvili, K. (2024b). AI creativity and human enhancement: The identity link. In F. Tigre Moura (Ed.), *Artificial intelligence, co-creation and creativity: The new frontier for innovation*. Routledge.
Khutsishvili, K., Pavicic, N., & Combé, M. (2024). The challenge of co-creation: How to connect technologies and communities in an ethical way. In J. Pelegrín-Borondo, M. Arias-Oliva, K. Murata, & M.S. Romero (Eds.), *The leading role of smart ethics in the digital world*. https://dialnet.unirioja.es/servlet/articulo?codigo=9537215
Open and Agile Smart Cities and Communities (OASC). https://oascities.org
Rayna, T., & Striukova, L. (2015). Open innovation 2.0: Is co-creation the ultimate challenge? *International Journal of Technology Management*, *69*(1), 38–53.
Rayna, T., Striukova, L., & Darlington, J. (2015). Co-creation and user innovation: The role of online 3D printing platforms. *Journal of Engineering and Technology Management*, *37*, 90–102.
Resilient Cities Network. (2013). https://resilientcitiesnetwork.org
Roser, T., DeFillippi, R., & Samson, A. (2013). Managing your co-creation mix: co-creation ventures in distinctive contexts. *European Business Review*, *25*(1), 20–41.
Smart Cities and Communities. (2022). Shaping Europe's digital future. https://digital-strategy.ec.europa.eu/en/policies/smart-cities-and-communities
Spuzic, S., Narayanan, R., Abhary, K., Adriansen, H. K., Pignata, S., Uzunovic, F., & Guang, X. (2016). The synergy of creativity and critical thinking in engineering design: The role of interdisciplinary augmentation and the fine arts. *Technology in Society*, *45*, 1–7.
Taibi, M., & Ozolins, U. (2016). *Community translation*. London: Bloomsbury Academic.

Taylor, C. (1989). *Sources of the self: The making of the modern identity.* Harvard University Press.
Taylor, C. (1992). *The politics of recognition.* http://elplandehiram.org/documentos/JoustingNYC/Politics_of_Recognition.pdf
The International Federation of Consulting Engineers (FIDIC). (2021). *Building sustainable communities in post-Covid world (SOW 2021: State of the world).* https://fidic.org/sites/default/files/SOTW_2020_report04_FINAL.pdf

4

SMART AND CONNECTED COMMUNITIES

From top-down to bottom-up approaches to city development

Brittany Haemi Lee

Introduction

We are in a city development boom: in the past decade alone, 91 new cities have been announced (*The Economist*, 2024) in the developed and developing world. This unprecedented growth gives us pause to examine two factors: how to plan and execute these cities and where to build them. Historically, two processes to city development have been at the forefront: the top-down approach and the bottom-up approach. The top-down approach refers to one where the most powerful level of government makes land use decisions on behalf of citizens, and a bottom-up approach is the opposite, where the instigators are the citizens themselves. The rise of technological innovation and adoption has not only augmented each approach but also created a third type, an Integrated approach, a hybrid of the top-down approach and bottom-up where in theory, there is a balance between government and citizen needs and wants. As to where these new cities are being built, we will argue that although every city is different, with varying population density and technological index, it is ultimately the process (or how) that dictates and therefore trumps any pre-existing factors to becoming a smart and connected community.

In this chapter, we will examine and evaluate three different approaches to modern city development. First, we will define what a smart and connected community means today with the recent acceleration of technology. Then we will examine the first approach, top-down, by defining this approach and assessing the benefits (e.g. energy efficiency) and risks (e.g. cybersecurity attacks) through case studies in the Kingdom of Saudi Arabia, China, Japan, and South Korea. Next, we will examine the second approach, the bottom-up

DOI: 10.4324/9781003439325-6

approach by defining the approach and assessing its benefits (e.g. local innovation) and risks (e.g. challenges to scale) through case studies in Canada, Spain, and the European Union. Finally, we will examine the Integrated approach by highlighting key technologies that have accelerated a more cohesive and iterative feedback process between government and citizens (e.g. extended reality) through case studies in Estonia, the United States, and Finland.

Finally, we will evaluate the approaches by comparing two characteristics of each case study: population density and technological index. We hypothesize that a successful smart and connected community is developed in a city with pre-existing characteristics of high population density and high technological index to be conducive to any of the three approaches.

Methodology

To assess the three city development approaches (top-down, bottom-up, integrated), we will examine them against eight cities of different countries. Each of these countries differs in population density and technological index. Population density is defined by population per square kilometer of surface area and the technological index is determined by technological progress (e.g. price of electric vehicles) and technological adoption (e.g. broadband range). Table 4.1 shows a summary of the countries and their population density and technological index by order of appearance.

TABLE 4.1 Population density and technological index of countries

Country	*City case study*	*Approach*	*Population density (people per sq. km.) as of 2023*[a]	*2022 Technological index*[b]
Saudi Arabia	NEOM	Top-down	17	33
China	Guangdong	Top-down	150	55
Japan	Fujisawa	Top-down	345	54
Democratic People's Republic of Korea	Songdo	Top-down	216	58
Canada	Toronto	Bottom-up	4	51
Spain	Barcelona	Bottom-up	95	45
Estonia	Tallinn	Integrated	31	50
Finland	Helsinki	Integrated	18	57
United States	New Rochelle	Integrated	36	62

[a] *World Development Indicators* (World Bank Group, 2024).
[b] *Global Innovation Index* (World Intellectual Property Organization, 2022).

Smart and connected community

The foundation of the smart and connected community is technology but more so the level of integration said technology has into the city's fabric. As we fitfully embrace the Fourth Industrial Revolution, the question no longer becomes "if" but "with and how best" to implement technology in our city fabric in a complex and dynamic world. The lines between the physical, digital, and biological worlds are blurred (World Economic Forum, 2016).

Recent technological advancements have accelerated the types of technology systems in cities. Two key categories of technologies have made the most profound impact: the Internet of Things and the Smart Grid. Having real-time data for Informational Technology and Operational Technology (IT/OT) for static and mobile assets and infrastructure is becoming common. This extends into the utility network where energy distribution is maximized among infrastructure as a connected system. Progress like this comes with its own risk namely with data privacy and cybersecurity.

A near-equal amount of uncertainty can counter the rate of technological implementation due to novelty leading to apprehension and, in some instances, mistrust. Platforms for open data and shared technological resources bring scale and network effects; however, data privacy, especially around personal information and consumer data, is top of mind when dealing with data collected in public spaces. Data surveillance, the practice of monitoring and collecting data often without the person's knowledge (Zuboff, 2023), is increasingly becoming a factor when cities consider implementing technologies. For example, facial recognition capabilities have been a factor of contention when deploying surveillance cameras in public spaces which must balance the ethical scale of attacking privacy versus enabling safety.

Closely related is the unfortunate reality of increasing cybersecurity attacks. Even with the proper risk management and governance measures, everyone from large enterprises and small-to-medium-sized enterprises (SME) to municipal governments and civil society organizations are subject to an attack.

The smart and connected community is one where the technological system can enable a better citizen well-being compared to one without technological systems. It must augment, not accelerate in an isolated manner, city development.

Top-down and bottom-up approaches to city development

The ancient cities of Pompeii (Italy) and Alexandria (ancient Egypt) were absolute in two key areas that made them possible and significant: leadership and land. A dictator yielding all political power and swathes of undeveloped land made each ripe for success. However, this is rarely the case today. Land is a commodity almost globally defined by constrained supply and growing demand; it continues to be a limiting factor for our increasingly urbanized world. With urban areas already home to 55% of the world's population, this figure is expected to grow to 68% by 2050 (UN-Habitat, 2020). Land is so scarce and valuable now that society has "created" net new land by way of man-land islands, as first trailblazed by Dubai with the Palm Islands and set a precedent for many Middle Eastern and especially Asian nations with extremely high-density urban areas. While population is growing, the supply of land is not. Thus, it becomes a conversation of scarcity and effective allocation of resources. How can we best build our cities that impact the most people but in an effective way?

Top-down

The top-down approach to city development can be defined as one where government initiates the process for planning and developing land use. The land is analyzed, zoned, and evaluated to maximize its highest and best use whereby the financial gains based on tenancy will benefit in the long-term after development. This process excludes neighborhood effects, a factor we will explore in the next section.

Top-down city development needs three main factors to be successful: policy and regulation, infrastructure development, and public-private partnerships. Policy and regulation are where the government has created a favorable environment for the proper policy standards and frameworks to be produced for successful city development execution. When the government is fully on board, execution comes naturally. Take the Kingdom of Saudi Arabia where in 2017 its then-new deputy prime minister, Mohammed bin Salman, was given the go-ahead from the King to implement Vision 2030 (Habibi, 2019). Vision 2030 is a comprehensive long-term plan for transforming the economy, culture, and social institutions of Saudi Arabia aspiring to transform the nation to "a vibrant society", "a thriving economy", and "an ambitious nation" (Kingdom of Saudi Arabia, n.d.). One key program is leveraging its sovereign wealth fund, the Public Investment Fund, to fund and manage its "giga-projects". The giga-projects speak to the unprecedented large scale of the projects, with the largest one being NEOM, an economic city in the northwest corner of the Kingdom's Red Sea coast. This project touts it will be a world leader in everything from the "internet of things", "cognitive cities", and "autonomous vehicles across its five regions, including its most infamous for its architectural marvel and unprecedented nature, The Line (NEOM, n.d.).

In China, a similar, more gradual approach has been taken with the Chinese Communist Party's Five-Year Plans. These five-year plans of national priorities are translated into policy directives that trickle down to virtually any state entity (Heikkila & Xu, 2014). For example, the 11th five-year plan enacted between 2006 and 2010, in response to the preceding rapid urbanization; several clusters were developed to plan for this growth: Liuzhou as an economic development cluster; Zhuzhou as an urban expansion cluster; and Chengde and public-private partnerships cluster (Heikkila & Xu, 2014). Fast-forward to today, we see in China's current Five-Year Plan, the 14th covering 2021–2025, there are policies around improving the new urbanization strategy and the quality of urbanization development. Namely, there is explicit mention of the "integrated development of urban agglomerations" using the "two horizontals and three verticals" approach in agglomerations such as Guangdong, Chengdu-Chongqing, and Central Yunnan (Center for Security and Emerging Technology, 2021).

In both the Kingdom of Saudi Arabia and China, we see that government policies and regulations, standards and frameworks stemming from directives from higher levels of government have not only galvanized smart and connected community development but have also enabled it.

The second factor needed is infrastructure development. We will focus our attention on investments in technology including the Internet of Things and Smart Grids. For the modern city, data for operational efficiency and citizen well-being is paramount and is usually delivered through real-time data. This real-time data can be produced by sensors (e.g. radio-frequency identification (RFID) and global positioning system (GPS)) to infrastructure from roads and buildings to fleets and trains, to gather urban data and monitor, manage, and optimize workflows. The rise of interconnectedness, networking, and information exchange has opened many opportunities for cities. In a top-down approach to city development, this can be seen through the government partnering with an Integrated Communications Provider (ICP) to build connectivity in areas where the internet is needed for sensor communication. In a top-down approach, it is usually public infrastructure and services that can be applied such as Smart Grids, Smart Traffic Controls, Smart Utility Meters, and Smart Waste Management.

Complex technological systems usually need public-private partnerships too. This collaboration between the government and private sector is critical to secure funding and allocate resources properly. Japan has been a pioneer in smart and connected community development for decades in part as a result of its advancements in information technology but also its recognition and emphasis on its human-centricity (Nagumo, 2022).

A prime example of a successful smart and connected community that has significant infrastructure development in technology and a public-private partnership is Fujisawa Sustainable Smart Town in Fujisawa City, Kanagawa Prefecture. The premise for development is on resident lifestyles comprised of community, mobility, energy, security, and wellness, which then feed the design of smart spaces, and only after that foundation has been laid, the creation of smart infrastructure (Fujisawa SST, n.d.). The human comes first, and the technology follows.

We have analyzed three case studies of smart and connected communities from the Middle East, China, and Japan and have seen that there are material benefits from a clear and unified vision from policy, economies of scale from technological and human resources, and thus ease of implementation. However, there can be risks to this approach, especially at scale such as oversight from local engagement, bureaucratic inefficiency, and lack of community needs met.

Songdo in South Korea is a prime example of a failed smart city due to its complete lack of community feedback from its local stakeholders and relying too much on the technology and policy stewardship of the government. Songdo Information New Town is a new project initiated in 2001 by the South Korean government. Though a 21st-century project, its story began in 1979 when the idea was launched and the urban plan was drawn several years after (Dias, 2022) Songdo is a ubiquitous eco-city, or u-eco-city, with a mission to create a "Green Korea" (Shwayri, 2013). For its time, it was the most progressive and large-scale greenfield-based smart city project in the world with housing for 65,000 residents and 300,000 workers. A key pillar in the Songdo plan was technology. It featured ubiquitous broadband internet connections, a robust public transit system, and a water recycling system. Furthermore, it housed many international and Korean research and development companies to become a hub for IT and research in bio and nanotechnology. However, despite the technological fanfare, there was a distinct lack of involvement of all stakeholders, especially local communities. The focus of technology neglected socio-cultured infrastructure (Yigitcanlar, 2019) and may have cost Songdo its potential.

Bottom-up

The bottom-up approach to city development is motivated by the practical problems confronted by the urban areas and their residents (Tang, 2019). In this converse approach to the top-down, citizens are at the core. Change happens from smaller scales such as the neighborhood level, micro-clusters, and community organizations. The Quadruple Helix Model of Collaboration outlines businesses, civil society, academia, and public authorities as other stakeholders that can affect city development congruence with the citizen (Mora, Deakin, & Reid, 2018).

The bottom-up city development approach needs three main factors to succeed: community engagement, local innovation, and collaborative networks. Community engagement is critical as the bottom-up approach relies on the community as its instigator. A prime example of this is the impact of the Covid-19 pandemic on retail, namely the hospitality sector. In Toronto, like many other cities, restaurants with only indoor seating were severely impacted by the pandemic's spacing restrictions. Thus, those restaurants fortunate to be adjacent to a wider street soon had an opportunity to benefit from a creative initiative borne out of necessity. Due to the overwhelming demand of patrons and the extreme willingness to keep business running during the financial strains of the pandemic, the city responded with a policy to legalize outdoor dining on certain sidewalks and streets. In November 2021, the city of Toronto made the CaféTO program permanent and continues to waive permit fees (Toronto, 2021). Although the initial push was borne from a federal health policy (pandemic spacing restrictions), the result was instigated by citizens, which was itself the catalyst for municipal policy.

The second factor for successful bottom-up city development, local innovation, usually refers to technological innovation. Barcelona shows some great examples in the infrastructure and open data arenas (Capdevila & Zarlenga, 2015). In infrastructure, Guifi.net is an initiative created by citizens and has become the biggest free network in the world. In open data, the Smart Citizen Project is an open-source platform that allows individuals to easily collect and share data about different environmental variables in their city. In both examples, the instigator is the citizen, who leverages technology for urban change and does not rely on any policy for this impact.

Alongside community engagement and local innovation, collaborative networks tie the stakeholders together. The CLEVER Cities project is a community-driven and global initiative funded by the European Union to increase the knowledge of nature-based solutions, contribute data to EU policymaking, and ultimately promote the uptake of nature-based solutions globally (Cities, n.d.). The project's initial bottom-up approach of a loose network of over 34 stakeholders allowed it to dynamically become a slightly more formal structure by way of co-governance (Bradley, Mahmoud, & Arlati, 2022).

When all three factors of the bottom-up approach are present, the result is a city with community engagement, flexibility, and ripe for local innovation. However, the bottom-up approach may not always work due to a lack of cohesion and unified vision, challenges to scale, and resource constraints.

Examining an integrated approach

The approach to city development may not and usually is not as binary as deciding on a top-down approach versus a bottom-up one. Many times, cities take an integrated approach leveraging the strengths of both approaches resulting in collaborative frameworks and shared responsibilities. The key components of this integrated approach are inclusive policymaking, joint funding mechanisms, and technology.

Inclusive policymaking in Barcelona is a result of a hybrid partnership between public administration and civil society organizations creating a bottom-linked approach (Garcia, 2015) This partnership has spawned economic impacts in creative clusters and technology for urban regeneration. Further to inclusive policymaking, the need for funding mechanisms is critical. Amsterdam is a prime example of this with its Amsterdam Smart City. It eventually created a public-private partnership portal to incur co-funding from the private sector (Noori, 2020).

Leveraging technology

We will examine two streams regarding technology's impact on city development: data sharing and services platforms. For a data-sharing example, we look to Estonia, a pioneer in digital government, to see how the public sector can be the hub for all stakeholders through technology for a smart and connected community. At the coattails of its independence from the Soviet Union, Estonia sought to build a minimal and efficient state and by 2016 had one of the world's highest internet penetrations (Margetts & Naumann, 2017). Three main layers create this government as a platform model: X-Road, the infrastructure for secure data exchange between residents, businesses, and public institutions; the Electronic Identification (EID); and the eServices portal, the service layer where citizens can access over 800 services. This virtual, block-chained, and secure government (Heller, 2017) has dubbed Estonia as a pioneer and role model for how we build ICT for future cities.

For services platforms, we will examine the plethora of community engagement platforms that aim to bridge the gap between government and citizens. These platforms offer a digital means for citizens to share their thoughts and the government to collect their feedback in real-time toward data-driven decision-making. Moreover, it breaks the chasm between the two and offers a neutral space for each party to share insights and hear concerns. Granicus is a trusted partner to over 6000 government organizations, offering solutions ranging from digital communication and engagement to government experience cloud. Its OpenCities product offers a low-code, no-code web solution to respond to resident needs and empower the community (Granicus, 2024). Some have a hyperfocus on community engagement, like Social Pinpoint, which since 2013 has created over 40 tools to streamline engagement, mapping, and ideas from citizens on public and private projects in their city.

While these service platforms offer a conduit for citizen's voices to be heard, sometimes visualizing something and capturing the automatic response is more valuable. The advent and rapid acceleration of extended reality (XR) which encompasses virtual reality (VR), mixed reality, and augmented reality, has offered a plethora of opportunities for participatory planning that at its core is an agile process that allows both top-down policy and bottom-up feedback to work in synchronicity. In the city of New Rochelle, the Department of Development created NRVR, its community engagement platform that utilizes VR headsets for citizens to visualize and provide feedback to in-progress projects in the design stage (EngageNR, 2024).

However, at its current stage, XR is dependent on the hardware that it functions on, for example., cost of a VR headset or large image files to render visualizations at scale. Thus, the rise of WebXR is particularly appealing for participatory planning of spatial planning due to its agile and scalable nature (Rzeszewski & Orylski, 2021). When combined with geographic information systems (GIS), such powerful use cases include an immersive multi-sensory experience that can induce emotional responses; the ability to compare existing and planned buildings simultaneously; and better visualization of urban elements such as air quality.

A web-based architecture is also what the city of Helsinki identified as the solution to its problem of disconnect between public feedback and final decisions. It was not able to capture the true representation of all groups. Thus, it partnered with UrbanistAI to develop a cloud-based platform that has two distinctive features: the ability to train the AI algorithm with local policy requirements and its AI-powered drawing and rendering tools for co-design (UrbanistAI, 2024). The resulting UrbanistAI platform was implemented in 2023, and its success from the local Summer Streets initiative received acclaim and resulted in adoption from other European cities in the Netherlands, Finland, and Germany (European Commission, 2024).

The integrated approach of city development allows for a balanced approach of planning and locals and improved scalability and adaptability. When it goes awry, there can be a conflict of interests and policies between stakeholders.

Discussion

In this chapter, we explored three approaches to city development to determine if there is a one-size-fits-all approach to planning and executing modern and future smart and connected communities. We used several case studies with differing population densities and technological indices to examine these three approaches. We hypothesized that a smart and connected community needs both a high population density and a high technological index. We found that neither of these was needed for any of the three approaches to be successful.

Looking at the countries with the highest population density in Table 4.2, Japan and South Korea, we see that both of these countries have a similar technological index range to Estonia and Finland. Estonia and Finland not only have dramatically lower population densities than Japan and South

TABLE 4.2 Population density and technological index of countries by descending population density

Country	*City case study*	*Approach*	*Population density (people per sq. km.) as of 2023*[a]	*2022 Technological index*[b]
Japan	Fujisawa	Top-down	345	54
Democratic People's Republic of Korea	Songdo	Top-down	216	58
China	Guangdong	Top-down	150	55
Spain	Barcelona	Bottom-up	95	45
United States	New Rochelle	Integrated	36	62
Estonia	Tallinn	Integrated	31	50
Finland	Helsinki	Integrated	18	57
Saudi Arabia	NEOM	Top-down	17	33
Canada	Toronto	Bottom-up	4	51

[a] *World Development Indicators* (World Bank Group, 2024).
[b] *Global Innovation Index* (World Intellectual Property Organization, 2022).

Korea (at 31 and 18 respectively) but also have a dissimilar city development approach (integrated).

When we look at the country with the lowest technological index in Table 4.3, Saudi Arabia, we see that it also has the second-lowest population density; however, it is responsible for one of modern day's most ambitious city development plans.

Through both Tables 4.2 and 4.3, we see that neither population density nor the technological index of the country is an indicative characteristic of that city's approach.

Conclusion

The development of the smart and connected community is more complex than ever before. The Fourth Industrial Revolution has redefined how we approach technology for our modern and future cities. The smart and connected city is one where the technological system can enable better citizen well-being as compared to one without technological systems.

There are three main methods to city development: top-down, bottom-up, and an integrated approach with the two. A top-down approach is where government officials dictate land use with little to no input from the citizens. To be successful, this approach needs robust policy and regulation, infrastructure development, and public-private partnerships. Examining three

TABLE 4.3 Population density and technological index of countries by ascending technological index

Country	*City case study*	*Approach*	*Population density (people per sq. km.) as of 2023*[a]	*2022 Technological index*[b]
Saudi Arabia	NEOM	Top-down	17	33
Spain	Barcelona	Bottom-up	95	45
Estonia	Tallinn	Integrated	31	50
Canada	Toronto	Bottom-up	4	51
Japan	Fujisawa	Top-down	345	54
China	Guangdong	Top-down	150	55
Finland	Helsinki	Integrated	18	57
Democratic People's Republic of Korea	Songdo	Top-down	216	58
United States	New Rochelle	Integrated	36	62

[a] *World Development Indicators* (World Bank Group, 2024).
[b] *Global Innovation Index* (World Intellectual Property Organization, 2022).

examples of top-down approaches gives us the perspective of the size and scope: Kingdom of Saudi Arabia for a modern example with its ambitious plans through its five giga-projects; China with the development of economic and technological hubs as a result of its most recent Five-Year Plans; and Japan for how infrastructure government through public-private partnerships can be achieved. South Korea's Songdo is a prime example of how plans can sometimes fail even with a firm goal in mind.

A bottom-up approach is one where city development is initiated from citizen's needs and then impacts policy and decision-making. To be successful this approach needs community engagement, local innovation, and collaborative networks; the Canadian city of Toronto's CaféTO project and the Spanish city of Barcelona's CLEVERCities project are excellent examples of the bottom-up approach.

An integrated approach of the two is highly dependent on leveraging technology namely data sharing and services platforms. Estonia is a pioneer in data sharing with its government as a platform model that allows an integrated, secure, and fluid system for citizen and government engagement. Service platforms such as Granicus enable municipal governments to leverage technology for community engagement. Furthermore, the advent and rise of XR offers an opportunity to visualize and respond in real time as seen with the projects with NRVR in New Rochelle and UrbanistAI in Helsinki.

The process toward a smart and connected city is complex. There is no one way to achieve it; it is within the complexities that bear opportunity for a successful city development.

Bibliography

Bradley, S., Mahmoud, I. H., & Arlati, A. (2022). Integrated collaborative governance approaches towards urban transformation: Experiences from the CLEVER cities project. *Sustainability*, *14*(23, 3–4.

Capdevila, I., & Zarlenga, M. I. (2015). Smart city or smart citizens? The Barcelona case. *Journal of Strategy and Management*, *8*, 266–282.

Center for Security and Emerging Technology. (2021, May 13). *Outline of the People's Republic of China 14th five-year plan for national economic and social development and long-range objectives for 2035*. Center for Secuirty and Emerging Technology. https://cset.georgetown.edu/publication/china-14th-five-year-plan/

Cities, C. (n.d.). *About the project*. CLEVER Cities. https://clevercities.eu/the-project/

Dias, S. (2022, February 8). *Songdo: The story and the history of a dream*. The Korea Times – National. https://www.koreatimes.co.kr/www/nation/2022/02/113_322915.html

Documentary, V. (Director). (2019). *Shoshana Zuboff on surveillance capitalism* [Motion Picture].

EngageNR. (2024). *Welcome to EngageNR!* https://engagenr.com/

European Commission. (2024). *Reimagining Helsinki: Participatory urban planning with Generative AI*. Public Sector Tech Watch. https://joinup.ec.europa.eu/collection/public-sector-tech-watch/reimagining-helsinki-participatory-urban-planning-generative-ai

Fujisawa SST. (n.d.). *About Fujisawa SST*. Fujisawa SST. https://fujisawasst.com/EN/project/service/

Garcia, M. (2015). Social innovation and creativity in cities: A socially inclusive governance approach in two peripheral spaces of Barcelona. *City, Culture, and Society*, *6*, 93–100.

Granicus. (2024). *Build meaningful connections with every digital ineraction*. Granicus. https://granicus.com/

Habibi, N. (2019). *Implementing Saudi Arabia's vision 2030: An interim balance sheet*. Waltham, MA: Crown Center for Middle East Studies.

Heikkila, E., & Xu, Y. (2014). Seven prototypical Chinese cities. *Urban Studies*, *51*, 827–847.

Heller, N. (2017, December 18). *Estonia, the digital republic. The New Yorker*. https://www.newyorker.com/magazine/2017/12/18/estonia-the-digital-republic

Kingdom of Saudi Arabia. (n.d.). *Vision realization programs*. Vision 2030. https://www.vision2030.gov.sa/ar/explore/programs

Margetts, H., & Naumann, A. (2017, February 28). *Government as a platform: What can Estonia show the world?*. Centre for Technology & Global Affairs. https://www.ctga.ox.ac.uk/article/government-platform-what-can-estonia-show-world

Mora, L., Deakin, M., & Reid, A. (2018). Strategic principles for smart city development: A multiple case study. *Technological Forecasting & Social Change*, *142*, 70–97. doi:10.1016/j.techfore.2018.07.035

Nagumo, T. (2022). *Measuring and monitoring liveability and well-being for smart cities.* Tokyo: Smart City Institute Japan.
NEOM. (n.d.). *Changing the future of design & construction.* NEOM. https://www.neom.com/en-us/our-business/sectors/design-and-construction
Noori, N. (2020). Classifying pathways for smart city development: Comparing design, governance and implementation in Amsterdam, Barcelona, Dubai, and Abu Dhabi. *Sustainability*, *12*(10), 8–13.
Rzeszewski, M., & Orylski, M. (2021). Usability of WebXR visualizations in urban planning. *International Journal of Geo-Information*, *10*(11), 2–3.
Shwayri, S. T. (2013). A model Korean ubiquitous eco-city? The politics of making Songdo. *Journal of Urban Technology*, *20*, 39–55.
Tang, Z. E. (2019). Identifying smart city archetypes from the bottom up: A content analysis of municipal plans. *Telecommunications Policy*, *43*, 1.
The Economist. (2024, Mar 7). *The world is in the midst of a city-building boom. The Economist.* https://www.economist.com/finance-and-economics/2024/03/07/the-world-is-in-the-midst-of-a-city-building-boom
Toronto, C. (2021, November 10). *Toronto City Council makes CaféTO permanent to support local restaurant industry.* Toronto. https://www.toronto.ca/news/toronto.city-council-makes-cafeto-permanent-to-support-local-restaurant-industry/
UN-Habitat. (2020). *World cities report 2020.* Nairobi, Kenya: United Nations Human Settlements Programme (UN-Habitat). Retrieved from The Value of Sustainable Urbanization.
UrbanistAI. (2024). *Shifting stakeholders' roles from commentators to change-makers.* UrbanistAI. https://site.urbanistai.com/
World Bank Group. (2024, June 28). *WV.1 World Development Indicators: Size of the economy.* World Bank Group. https://wdi.worldba1*nk.org/table/WV.1?1?_gl=1*1*nzd4ry*1?_gcl1?_au*MjA0ODM5MzY4Mi4xNzI0NzY5MzA4#
World Economic Forum. (2016, Jan 14). *The Fourth Industrial Revolution: What it means, how to respond.* https://www.weforum.org/agenda/2016/01/the-fourth-industrial-revolution-what-it-means-and-how-to-respond/
World Intellectual Property Organization. (2022). *Global innovation index 2022.* Geneva: World Intellectual Property Organization.
Yigitcanlar, T. (2019). The making of smart cities: Are Songdo, Masdar, Amsterdam, San Francisco and Brisbane the best we could build? *Land Use Policy*, *88*, 4.
Zuboff, S. (2023). The age of surveillance capitalism. In *Social theory re-wired* (p. 11). Routledge.

PART II

Smart cities and governance

5

SMART CITIES AND PUBLIC DATA – CURRENT CHALLENGES AND POTENTIAL SOLUTIONS

Ana Rita Antunes, Mariana Silva, Bárbara Patrício, Margarida Prozil, and António Ferreira Dias

Introduction

In this era of transformative change, the concept of smart cities emerges as a dynamic fusion of technological advancements, organizational enhancements, and political innovations (Neves et al., 2020). Despite the abundance of articles and reports on smart cities, a universally accepted definition remains elusive. However, a common thread in most definitions emphasizes the use of Information and Communication Technologies (ICT) to enable city development, improve urban services, and enhance stakeholder access to information. Smart cities globally leverage city-wide investments in ICT to drive technological innovation, foster economic development, maintain environmental sustainability, and enhance the overall quality of life for citizens (Law & Lynch, 2019). Additionally, smart cities inspire forward-thinking and innovative approaches across governance, economy, environment, living conditions, people, and mobility-centric initiatives (Bibri & Krogstie, 2017). Table 5.1 depicts the areas and examples of datasets for each cluster of smart cities, based on ISO 37120. Published by the International Organization for Standardization (ISO), ISO 37120 (Midor & Płaza, 2020) provides a framework for cities to measure and compare their performance in various aspects. This standard aims to enhance the consistency and comparability of data across cities, fostering sustainable urban development and improving the overall quality of life for residents (Przybyłowski et al., 2022).

As the foundation of the Fourth Industrial Revolution lies in data, governments have responded by increasing openness and transparency, promoting the release of data in accessible formats, marking a global shift towards decision-making. This transformation presents a challenging yet opportune

DOI: 10.4324/9781003439325-8

TABLE 5.1 Smart city clusters and areas with respective examples of datasets

Cluster	*Area(s)*	*Examples of datasets*
Smart governance	Governance; telecommunication; urban planning	Administrative records; voter participation in elections; telecommunication infrastructure data; urban development plans.
Smart economy	Economy; finance	Economic indicators; GDP data; unemployment rate; financial transactions; tax revenues.
Smart environment	Environment and climate change; solid waste; energy; wastewater; water	Air quality measurements; greenhouse gas emissions; waste disposal records; water and energy usage statistics; energy production from renewable sources.
Smart living	Education; health; housing; recreation; safety; sports and culture; urban/local agriculture and food security	Educational performance metrics; healthcare outcomes; average life expectancy; housing affordability data; crime rates; cultural event attendance; agricultural production statistics.
Smart people	Population and social conditions	Demographic data; natality and mortality rates; social welfare statistics.
Smart mobility	Transportation	Traffic flow data; public transportation records; parking space occupancy; road infrastructure conditions.

moment to introduce innovative technologies, data-driven insights, and collaborative approaches that can catalyze positive social and economic transformations (Currie, 2020; S. Verhulst et al., 2018).

Public data encompasses information generated, collected, processed, and made accessible by governments or public institutions using, for example, reports. On the other hand, the open data concept has evolved to denote the collection of large sets of information that are not only freely available but also openly used, modified, and shared by citizens without restrictions (Capgemini Institute & European Commission, 2015). Exploring the various impacts and outcomes of open data initiatives reveals four primary dimensions. Firstly, as the global economy becomes increasingly reliant on information, the accessibility of open data can foster business creation, attract foreign investment, and lead to meaningful job generation, thereby contributing significantly to economic opportunity. Secondly, these initiatives involving open data play an essential role in allowing the resolution of complex public problems by providing the opportunity to improve situational awareness, increase knowledge and expertise, and enable targeted interventions by policymakers, civil society, and citizens. At the same time, it also facilitates impact tracking. Thirdly, open data initiatives lead to improved governance through enhanced transparency, accountability, efficient service delivery, and increased information sharing within and outside city domains. Lastly, open data empowers citizens by enhancing participatory capacity and acting as a catalyst for social mobilization (S. G. Verhulst & Young, 2017).

In alignment with the shifting dynamics of open data, FIWARE, a European Commission–sponsored project in collaboration with the ICT sector, focuses on developing technologies for the Internet of Things (IoT) and the future Internet. With the mission to establish an open and sustainable ecosystem around free software platform standards, FIWARE facilitates the creation of smart applications across sectors. By providing Application Programming Interfaces (APIs) based on open specifications, FIWARE introduces generic enablers and key components covering security, data management, cloud hosting, and device management. These enablers serve as essential building blocks for diverse smart applications, offering a comprehensive framework for innovation (Araujo et al., 2019). Additionally, as the digital environment continues to evolve, initiatives similar to Gaia-X play a crucial role in addressing concerns related to industrial data control when hosted by non-European cloud service providers. Launched in 2019 by the German and French Ministers of Economy and led by a collaborative association, the Gaia-X project aims to promote the adoption of cloud services and to facilitate data exchange among European businesses. In a complementary effort to FIWARE, Gaia-X strategically assists industries in defining tailored data spaces and developing compelling use cases. Subsequently, these can then be deployed on a compliant-by-design platform architecture, aligning with Gaia-X specifications, trust principles, and labelling frameworks (Tardieu, 2022).

Data management and associated challenges

In today's fast-paced business world, it has never been more important to make smart decisions to accompany the dynamic and evolving technological landscape. By allowing organizations to identify trends and patterns, data analysis has revolutionized the decision-making processes in a data-driven approach. Given the exponentially growing amount of data, it is not just about having large volumes of information, but it is about using it wisely. Moreover, as decisions become increasingly complex and multifaceted, data analysis emerges as a strategic asset, offering more than raw information and an opportunity to optimize processes, operations, and resources (Provost & Fawcett, 2013). The true value of data lies not just in its abundance but in its insightful examination. Therefore, the process of data analysis involves the uncovering of hidden patterns, identification of trends, and extraction of meaningful and actionable insights, constituting the essence of this transformation. In a world saturated with information, the ability to analyse data assumes notable importance for informed decision-making, guiding businesses towards more effective strategies and fostering innovation (John Walker, 2014).

The flow of information within the data life cycle

As a series of phases over the course of its usefulness, information flows through a lifecycle of different stages, commonly referred to as data life cycle (Figure 5.1). This systematic process encompasses key interconnected stages, including data generation, collection, processing, storage, management, analytics, and reporting (Wing, 2019).

Effectively managing this cycle requires a thorough understanding of each stage. By carefully handling all the steps of the data life cycle, organizations not only improve processes but also enhance overall efficiency. The first step, data generation, involves the continuous creation of information from diverse sensors, citizen interactions, and automated systems in real time (Rahul & Banyal, 2020). In the context of smart cities, the valuable contributions of open data platforms, along with data from government agencies, public services, and social media, significantly enrich datasets and contribute to a

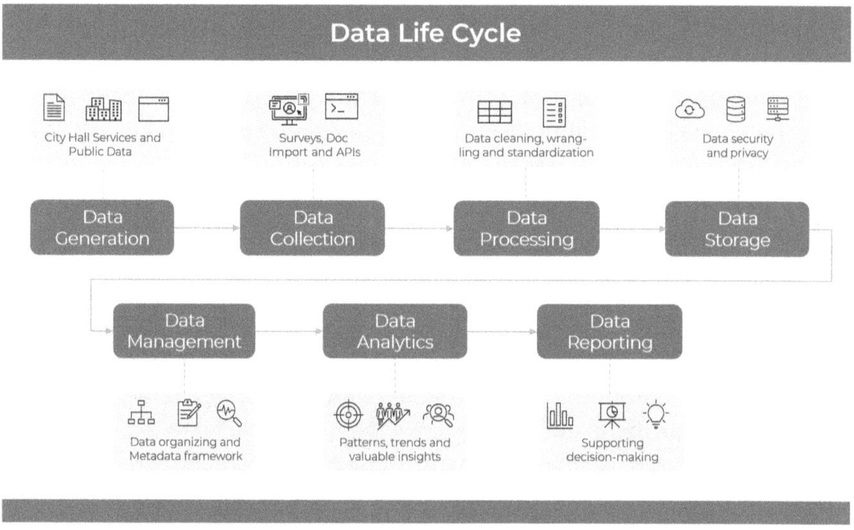

FIGURE 5.1 Representation of the data life cycle steps: data generation, data collection, data processing, data storage, data management, data analytics, and data reporting.

Abbreviation: APIs: Application Programming Interfaces.

comprehensive understanding of urban dynamics. The exploitation of all available data sources, with a particular emphasis on public data, is essential for smart cities to enhance efficiency and support informed decision-making (Neves et al., 2020).

Data collection is the subsequent strategic process that involves the systematic gathering of data from diverse sources, focusing not only on traditional methods, such as direct measurement and surveys, but also on leveraging modern technologies and platforms. Examples of these modern technologies and platforms include (i) mobile applications and geolocation data to analyse user preferences and movement patterns, (ii) IoT devices for collecting interconnected data from sensors and machines, (iii) social media data collection to understand trends and user behaviour, (iv) web scraping to extract information from websites and surveillance and security systems for video and audio data, and (v) satellite imagery and remote sensing for applications such as environmental monitoring and urban planning. This integration of traditional and modern methods increases the scope and efficiency of data collection across various domains. Therefore, open data platforms play a central role in this process, serving as hubs where information is made accessible to the public. Through APIs, smart cities can establish dynamic connections with these platforms, facilitating a seamless extraction of valuable data (Neves et al., 2020).

The next step in the cycle is data processing, which implies the refining of raw data into a structured and usable format. This stage initially focuses on meticulous data cleaning and wrangling as essential processes to rectify inconsistencies, eliminate errors, and standardize formats. In this way, it is ensured that subsequent stages in the data flow build upon a foundation of accurate and coherent information. Additionally, it is equally imperative to integrate robust data compression techniques and implement stringent data security measures (Wing, 2019). This dual approach serves a twofold purpose: first, to reduce the storage footprint while retaining essential information, addressing the escalating volume of data generated in smart cities; second, to safeguard the privacy and integrity of public data, fostering confidence in the reliability and trustworthiness of the smart city's data management practices (Braunschweig et al., 2012).

Furthermore, the data storage phase incorporates continuous quality assurance measures to uphold the integrity of stored information. This includes ongoing monitoring and validation processes, ensuring the accuracy and relevance of the data. Simultaneously, robust data security measures and privacy precautions are integrated into the storage infrastructure, including encryption techniques to protect against unauthorized access and uphold the privacy of public data. The establishment of stringent data access rules adds an additional layer of control, ensuring that only authorized entities can retrieve and utilize specific datasets. This approach within the data storage phase not only addresses the efficiency and scalability of storage systems but also underscores the commitment to data quality, security, and privacy (Zhang et al., 2022).

Moreover, the data management step is the process within the data life cycle that focuses on organizing, optimizing, and ensuring the quality of the gathered information. It involves creating and maintaining metadata frameworks for effective organization and searchability, adapting to the diverse nature of urban data. Furthermore, data management integrates the entire data life cycle seamlessly, aligning data generation, collection, processing, and storage. This phase also emphasizes continuous monitoring and improvement, upholding data quality, security, and privacy standards (Rahul & Banyal, 2020).

Additionally, data analytics is the next step that comprises the examination of data using statistical methods, machine learning algorithms, artificial intelligence, or other analytical techniques to uncover patterns, trends, and valuable insights. Essential for supporting decision-making, data analytics enables stakeholders to navigate urban complexities, fostering proactive responses and informed strategies (Sarker, 2021).

As the final step of the data life cycle, data reporting aims to efficiently share insights from data analytics with city authorities, residents, and local enterprises. By facilitating access to pertinent information, this communication promotes collaboration and empowers stakeholders to actively contribute to the enhancement of urban environments. In the context of smart cities, transparent and accessible data reporting becomes crucial for building trust and encouraging widespread involvement in ongoing developments (Attard et al., 2015).

Open data challenges

Taking into consideration the stages of the data life cycle, with a special focus on data analytics and its importance for business insights, it becomes imperative to navigate through the challenges posed by open data in the domain of smart cities. By understanding and addressing the obstacles inherent to leveraging open data initiatives in urban environments, one sets the stage for a guaranteed data quality, accuracy, reliability, and security, which is essential to maintain a robust data collection process (Neves et al., 2020). Such challenges include difficulties in addressing technological and human-related issues during data collection. For instance, data collected through sensors may have noise in the measurements, and manual entries (e.g., via Excel files) can introduce human errors. Furthermore, when multiple entities are assigned the responsibility of reporting the same indicator, it becomes crucial for each of them to actively provide the means to relate and analyse the information effectively, yielding more insightful conclusions (Attard et al., 2015). Additionally, data processing steps are pivotal to achieve data quality, which involve having a clear attribute semantics, no missing values, and machine-readable data. However, many datasets may not initially meet these criteria, requiring manual cleaning and transformation processes to rectify discrepancies and enhance the overall quality of the data (De Dutta & Prasad, 2020).

Another concern relates to data discoverability as an essential aspect of any data strategy since it involves providing ample metadata, such as descriptions and temporal information, to ease and streamline the process of finding datasets and data sources. This practice not only facilitates a comprehensive understanding of the data but also ensures clarity regarding the meaning of its features. However, the adoption of standard methods and their consistent implementation are commonly missing in many platforms, resulting in inadequate metadata that severely limits data discoverability. In the processing step, the absence of reliable metadata impedes the identification of inconsistencies and potential errors and hinders data quality assessment. These outcomes illustrate the importance of prioritizing metadata provision for effective data management and analysis (Braunschweig et al., 2012).

In parallel, addressing data fragmentation is another crucial point, as open data may be dispersed across multiple platforms and sources. This dispersion creates obstacles to easily access comprehensive datasets, and ultimately require users to navigate through different platforms to gather information. Besides that, the absence of uniformity across different datasets on open platforms may hinder integration and analysis. Aspects such as different data structures, divergent units of measurement, or distinct coding conventions may impose an additional effort in the process of data harmonization by adding complexity to the information aggregation tasks (Neves et al., 2020). Thereby, an effective data integration requires the data to be compatible and consistent across the different datasets (Braunschweig et al., 2012). The demand to make reported data accessible to different stakeholders adds another layer of complexity, which requires the implementation of user-friendly interfaces and clear presentation methods. The interoperability aspect and the integration of reporting systems with other platforms and datasets emerge as key features for a comprehensive understanding of the reported information. Conversely, a robust technical infrastructure is equally crucial to support efficient data reporting, as inadequate systems may result in delays, errors, or accessibility issues (Attard et al., 2015; Jaakkola et al., 2014).

A major challenge in the analysis of smart city data is keeping up with the evolving data skills, competencies, and technical barriers. The analysis of open data often requires a level of technical proficiency, and users without a strong background may find it challenging to navigate raw data or to apply the necessary processing steps for deriving meaningful insights. To overcome this issue, there is a growing emphasis on enhancing user skills through training programs and promoting the development of user-friendly analytical tools. Given the high demand and constant change of these capabilities, an investment in educational initiatives can empower professionals and the broader audience to expand data skills, fostering a more inclusive and collaborative approach to data analysis within real-world projects, challenges, and datasets, such as smart city initiatives (Lai et al., 2020; Williamson, 2015).

As mentioned previously, ensuring the accuracy and reliability of reported data is a primary concern, since data inaccuracies or inconsistencies may lead to misunderstandings in the identification of patterns. Thus, the challenge in the employment of public data for data analytics resides in the process of determining the value and impact of the insights generated. Indeed, reliable results are not guaranteed and, for that reason, it is important to think critically about the relevance, accuracy, and potential biases of public data (Moustaka et al., 2019).

Another challenge faced by smart cities involves data privacy and security. Considering the exacerbated generation of data observed nowadays, there is a growing need to protect information from unauthorized access and misuse, particularly for sensitive data like personal or financial data. Other significant concern are the data storage and management steps, since this information requires protection against tampering or leakage. Therefore, prioritizing robust measures for data protection is fundamental to maintain the integrity of open data in smart cities and build trust among stakeholders and citizens (Neves et al., 2020). Possible mitigation measures include the use of technologies such as blockchain, which offers a decentralized and distributed solution for enhancing data security and privacy. However, its implementation still faces technical and regulatory challenges (De Dutta & Prasad, 2020).

In addition, it is crucial to recognize that the programming languages and coding involved in the process of data collection can be inherently complex. The intricacies related with the design and implementation of data extraction processes demand careful consideration, especially when dealing with diverse data sources. The preprocessing steps applied to the data on these platforms may potentially limit the granularity or detail available for analysis. Consequently, ensuring the proficiency of the extraction program becomes vital for obtaining accurate and meaningful data. The complexity may arise from factors such as data formats, source variations, and evolving technologies, emphasizing the need for adept programming skills and thoughtful implementation strategies in the data extraction phase of the collection process (Pereira et al., 2017).

On top of this, handling large volumes of data also poses a considerable concern for smart cities. The escalating quantity of generated data introduces complexities, which can result in performance issues and increased operational costs (Neves et al., 2020). Besides that, scalability poses a practical challenge as the volume of data grows. Ensuring that systems can effectively handle the increasing volume of data is essential for maintaining the reliability and effectiveness of data analytics processes (Gharaibeh et al., 2017). Conversely, ensuring effective data management involves challenges of data timeliness. Many open datasets may not undergo frequent updates, potentially rendering them obsolete over time. Thus, outdated data hampers its relevance for current analyses and decision-making processes, emphasizing the importance of regular updates to maintain data accuracy and applicability (Neves et al., 2020).

Additionally, managing data ownership and attribution poses complexities, as determining the rightful owners and ensuring proper attribution for open data can be unclear. To navigate ethical and legal considerations associated with data usage, a clear understanding of the rights and responsibilities involved becomes crucial in the data management step. Clear guidelines and practices are necessary to foster responsible and transparent handling of open data in smart cities. Moreover, while community contributions on crowdsourcing platforms may offer benefits in promoting user engagement and obtaining feedback for data enhancement, nurturing an active and collaborative community presents its own set of challenges (Attard et al., 2015).

Data **warehouse** *as a centralization system*

Each stage of the data life cycle, from acquisition to processing, analysis, and dissemination, presents unique challenges. Some of these challenges include accessibility, consistency, quality, and cost-effectiveness of the processes, which can be addressed by data warehouses. At this point, the introduction of a robust data warehouse architecture emerges as a critical solution as it assumes an important role in managing extensive datasets from diverse sources for informed decision-making, as noted by Bilal et al. (2016). According to the definition of Souibgui et al., a data warehouse is a subject-oriented, non-volatile, integrated, time-variant collection of data designed to support management decisions (Souibgui et al., 2019). When we look at it from a different perspective, data warehousing is the systematic collection and arrangement of data from various sources with the goal of obtaining a comprehensive and detailed image of a business, whether by focusing on specific areas or capturing all its operations (Gardner, 1998).

In the context of smart cities, data warehouse serves as a centralized repository for storing, organizing, and retrieving vast amounts of data across sectors and services, enhancing data accessibility, reliability, and consistency (Bilal et al., 2016). To exemplify the system, Figure 5.2 illustrates a data warehouse architecture, consisting of four main layers, including data generation/sources; data integration and Extract, Transform, and Load (ETL) tools; data warehouse database for data storage; and query and analytical tools.

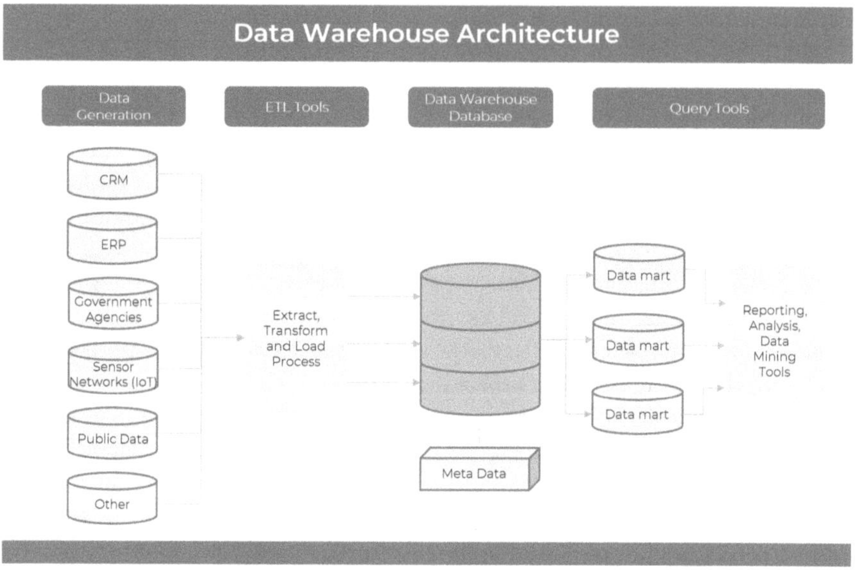

FIGURE 5.2 Data warehouse architecture and respective layers: data generation, ETL tools, data warehouse database; and query tools.

Abbreviations: CRM: Customer Relationship Management; ERP: Enterprise Resource Planning; ETL: Extract, Transform, and Load; IoT: Internet of Things.

In theoretical terms, data can be generated through various channels, including customer relationship management (CRM), enterprise resource planning (ERP), government agencies, sensor networks and IoT, public data, and other sources. The diversity of data is a crucial point to consider for the usefulness and efficiency of a data warehouse since incorporating data from a large variety of channels allows the data warehouse to provide a comprehensive and holistic view of all the operations (Watson, 2002).

Besides that, the central part of the establishment of the data warehouse system is the ETL process, described as a data refinement pipeline and ensuring that only high-quality, well-organized, structured data enters the data warehouse database. In its core, ETL tools facilitate the extraction of data from diverse sources, allowing its transformation into a standardized format, and loading the information into the data warehouse. This three-phased approach serves as a bridge, which connects raw and heterogeneous data from different sources to the structured and organized data warehouse. In this process, the following steps may be necessary to address the complexities inherent of open data (Kimball et al., 2004):

- Data cleaning: Identifying and rectifying errors, inconsistencies, and inaccuracies in the raw data and providing warnings in case of changes to prevent disruptions in subsequent stages. This step also involves these processes:

- Duplicate removal: Detecting and eliminating duplicate records to maintain data integrity.
- Outlier detection: Identifying and handling outliers or anomalous data points that may skew analysis results.
- Missing data handling: Addressing missing values through imputation, interpolation, or removal.
- Data type conversion: Converting data types to ensure compatibility and consistency across datasets.

- Data standardization: Ensuring that data adheres to a consistent format, unit, naming convention, or other predefined standards.
- Data validation: Verifying that data adheres to predefined rules, constraints, or validation criteria.
- Data transformation: Applying mathematical or logical transformations to derive new variables or metrics.
- Aggregation: Summarizing or aggregating data to different granularities for analysis or reporting.
- Data partitioning: Breaking down large datasets into smaller, manageable partitions for efficient processing.
- Data archiving: Moving older or less frequently used data to archival storage for optimization.
- Data versioning: Managing different versions of datasets to track changes over time.

Afterword, the data warehouse database serves as a robust foundation that efficiently stores and manages extensive datasets, providing a structured base for subsequent processes. In this layer, metadata plays an essential role as guardian of the context and description of information, which ensures a layer of abstraction that improves data understanding and contributes to the efficient management of the entire data ecosystem. The inclusion of metadata is valuable as it enables users to navigate and interpret the stored data, warranting clarity and facilitating effective utilization (Kimball et al., 2004).

Another layer of the architecture comprises query tools, which facilitate user empowerment by enabling interaction and insight derivation from data and provide a user-friendly query interface of the central data warehouse database, serving as a basis for business intelligence. Being seamlessly integrated with query tools, data marts further enhance this capability by providing specialized subsets of data and allowing users to access and analyse information tailored to their specific needs. Such combination of flexible query tools and precision-focused data marts enables organizations to make decisions more efficiently, providing users a full picture of the overall data situation and specific insights related to their areas of focus (Nambiar & Mundra, 2022; Runtuwene et al., 2018).

For the solutions built upon the data warehouse, it is important to adopt modelling best practices, such as the well-known star schema (Kimball et al., 2003). In this context, the organization of data into dimensions and fact tables improves the overall structure and accessibility of the data within many data warehousing systems. Dimensions represent the descriptive attributes or context of the data, acting to categorize and filter information, for example, time, geography, product, or customer. Fact tables, on the other hand, contain numerical measures or metrics such as sales, revenue, or quantities, and serve as the core of analytical processes (Bellatreche, 2012).

In summary, a data warehouse architecture offers many analytical advantages, since it is optimized for query performance, enabling complex analytical queries to be executed efficiently. Besides that, it facilitates the aggregation and summarization of data, making it easier to perform analyses at various levels of granularity, and to generate reports and dynamic presentations of data through dashboards. Furthermore, it allows organizations to analyse trends and changes over time by storing historical data, even when it is no longer available from the original sources. This capability becomes particularly valuable when APIs or other data interfaces are updated or modified, given the capacity of the data warehouse to retain historical information. Another advantage is the capacity for scalability, supporting the system to increasing volumes of data and user demands without sacrificing performance. Moreover, data warehouses minimize report maintenance since it is not necessary to modify it whenever the gathered information is updated. Lastly, data warehouses provide enhanced security and backup capabilities, making it simpler to control access and offering a secondary backup level in case of unforeseen incidents (Guerra & Andrews, 2011).

Territorial intelligence center: An open data centralization solution

From the knowledge gathered in the data life cycle and the challenges of using open data for smart cities, it is opportune to introduce a solution to make effective use of public data. Hence, the Centro de Inteligência Territoral[1] (CIT, as Territorial Intelligence Center), tailored to the context of Portugal, addressed the described challenges. With the capacity to assimilate data from various open platforms, this solution was conceptualized and developed to navigate each phase of data cycle, culminating in the creation of dynamic county reports. Its significance extends beyond local governance, reaching citizens, investors, and students, since it can act as a centralized repository for diverse public information. The architecture for the proposed solution for

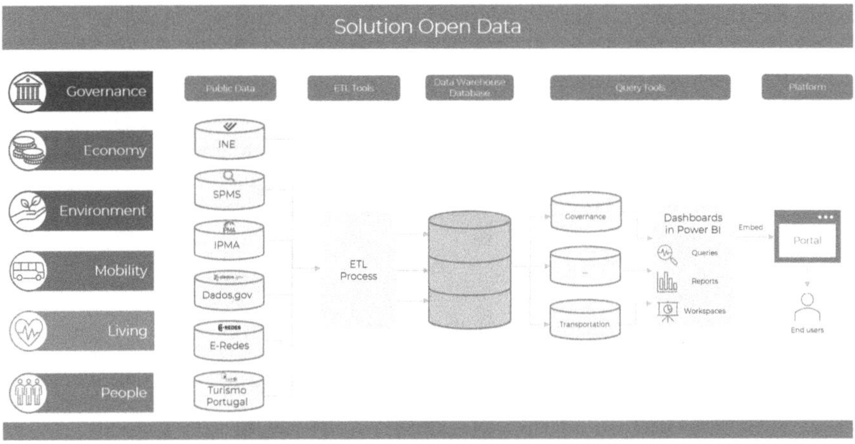

FIGURE 5.3 Architecture of the open data platform, comprising several open data sources in different areas, their integration and analysis steps, and final information visualization dashboards embedded into an accessible portal.

Abbreviations: INE: Instituto Nacional de Estatística (as Statistics National Institute); SPMS: Serviços Partilhados do Ministério da Saúde (as Shared Services of the Ministry of Health); IPMA Instituto Português do Mar e da Atmosfera (as Portuguese Institute for the Ocean and Atmosphere), ETL: Extract, Transform, and Load.

managing open data in the context of smart cities is represented in Figure 5.3. This intelligent system categorizes data based on distinct areas such as governance, economy, environment, mobility, living, and people (Table 5.1), providing stakeholders with a nuanced understanding of each facet. By improving accessibility and organization, the platform is positioned to transform how cities manage and present information. This not only enhances transparency and strategic insights but also contributes to more informed decision-making processes.

The first step in this platform solution was to connect to various open data sources, adeptly collecting data from diverse areas. Thereby, collection phase emerges as one of the most crucial aspects of such a project, requiring meticulous attention to the challenges previously mentioned. Considering the example of open data platforms in Portugal, notable entities include Instituto Nacional de Estatística[2] (INE, as Statistics National Institute), Serviços Partilhados do Ministério da Saúde[3] (SPMS, as Shared Services of the Ministry of Health), Instituto Português do Mar e da Atmosfera[4] (IPMA, as Portuguese Institute for the Ocean and Atmosphere), E-Redes[5] (related to energy distribution), Turismo Portugal[6] (related to tourism), Dados.Gov[7] (related to

TABLE 5.2 Portuguese public data platforms and respective areas of service

Open Data platform	*Main objective*	*Areas of service*
INE	Statistical data collection and analysis	Demographics, economics, and others
SPMS	Health data management and analysis	Healthcare
IPMA	Meteorological data provision	Weather, climate
E-redes	Energy distribution data management	Energy
Turismo Portugal	Tourism-related data provision	Tourism
Dados.Gov	Centralized repository of government datasets	Public administration; various government sectors

public administration), among others. The primary objectives of these platforms and the specific areas of service are described in Table 5.2.

In this phase, challenges emerged as these platforms exhibited issues such as incomplete or unavailable APIs, limited data retrieval in each query, and occasional lapses in updates, potentially resulting in drawbacks in dealing with obsolete data. When it comes to APIs, there was another frequent issue to overcome since some of them only provide recent data. To address this, it is essential to store historical data to conduct temporal analysis effectively. Another concern was the abundance of information in each source, requiring an extensive pre-analysis to filter out less relevant data. As an example, the INE platform contains more than 10.000 indicators, some resulting from census and updated every ten years, while others are updated yearly or monthly. In this scenario, careful assortment of data becomes vital, and factors such as update dates, relevance, periodicity, and geographic level (ranging from national to parish) play a decisive role. It is imperative to distinguish and prioritize indicators based on these criteria to ensure the extraction of meaningful and pertinent insights from the vast pool of available data.

Moreover, challenges also arose from data omission in certain municipalities. For instance, energy distribution operators differ based on geographic regions. Thus, the only way to have the country overview is if every energy supplier makes their data available. Otherwise, it becomes unattainable to achieve a holistic countrywide understanding of the subject. To prevent drawing incorrect conclusions, it is crucial to detect the causes of missing values by discerning between the actual zero value and a value being left blank or mistakenly assumed as zero. The lack of open information, particularly for crucial sectors in a city such as health and mobility, poses a substantial barrier to comprehensively understanding the prevailing conditions and dynamics across a territory.

The subsequent phase involved the integrity and robustness of the ETL process and the secure storage of data in a data warehouse database, such as MySQL, MariaDB, PostgreSQL, Redis, SQLite, among others (Olivares-Rojas et al., 2021). The selection of an appropriate database will depend on scalability, performance, and compatibility with the specific requirements of the smart city platform (Olivares-Rojas et al., 2021). Regarding query tools, the first phase implicated using programmatical languages such as Structured Query Language (SQL) or Python, as a practical means of analysis. The SQL queries were arranged into SQL views, acting as virtual tables to offer a dynamic approach and organize data without altering the original datasets. This functionality enables users to efficiently query and analyse data (Agrawal et al., 2000). During this phase, the focus laid on data integration and normalization by combining information from multiple sources to create a unified data mart.

A significant challenge resided in the inconsistent naming conventions present in different sources for similar dimensions or attributes, such as gender, age groups, and nationalities. Furthermore, difficulties in linking different datasets based on geographic location emerged, especially when municipality names are not consistently recorded, or when municipalities from different regions share the same name. To address this, a data standardization process was essential to map the diverse categorizations, mismatched units and scales, and temporal misalignment of data, prioritizing unique keys instead of names. Such a problem is many times accentuated by the absence of metadata. Additionally, different levels of granularity in datasets posed challenges to integration, which required aggregation/disaggregation to a common level. An inherent disadvantage of using open data was the occasional aggregation of information, complicating the linkage between datasets. For instance, datasets from INE present population data categorized by sex and by age in different datasets. This impedes the association of both dimensions to obtain data, for instance, on females under 20 years old. In another example, certain tax indicators cannot be derived from the available data, as they are provided by open data platforms as pre-calculated values for each municipality or region. Another problem arose from working with datasets categorized by different factors, for instance, the number of fires categorized by causes (negligence, natural, intentional, or undetermined). Occasionally, these datasets contain confidential or unspecified data in these categories, causing issues when the sum of the parts fails to match the total. To address this, a thorough review of the dataset was needed, possibly requiring the creation of newly unspecified categories to match the totals or their exclusion. Addressing these challenges systematically optimizes the integration process, resulting in a unified data mart that is accurate, coherent, and valuable for the subsequent stage.

Data visualization tools, such as Power BI, Tableau, and Qlik, are widely used for creating interactive and informative dashboards. These tools offer diverse visualization options and support dynamic data exploration, providing a user-friendly interface for report generation (Lousa et al., 2019). In Power BI, the semantic model is a key component that serves as a logical description of an analytical domain, containing metrics, business-friendly terminology, and representations that enable deeper analysis. It acts as a virtual layer between data sources and end users by defining relationships between facts and dimension tables, creating hierarchies, and establishing indicators. Furthermore, the semantic model contributes to data governance and security by implementing Object Level Security (OLS) or Row Level Security (RLS). This facilitates the establishment of access policies, restricting access to sensitive data (Box, 2022). Building on this, a notable feature for enhanced analysis is the Analyze in Excel tool, that creates an Excel PivotTable connected to the data hosted in Power BI (Becker & Gould, 2019). While dashboards present main results and key performance indicators, this tool allows users to navigate beyond the dashboard visuals by providing a more granular and dynamic data exploration. Such integration empowers users with the flexibility to perform in-depth analyses, complementing the summarized insights presented in the dashboard.

Following the creation of dashboards across the various areas, the next step involved embedding in a web portal, which can be a website that acts as the central access point for users, facilitating navigation, presenting data, and providing access to documents and reports. It serves as a bridge, ensuring the interconnection of all functional and technological components while implementing access security protocols, monitoring, and customizing information. This centralized approach allows for the insight aggregation in one location, offering users a cohesive and secure exploitation of analytical outputs.

Building on this concept, the presented architecture, denominated as CIT, was developed, and implemented in Portugal. The solution consists of a tool that continuously aggregates countrywide open data, which is collected from the available data source platforms indicated in Figure 5.3 for the different city areas. Additionally, CIT foresees scalability and is extensible to internal municipality data gathered by each region. The portal, accessed in a log-in based manner, contains personalized information panels with all information from municipalities, further compared by respective region and by country. Thus, municipalities can analyse their data effortlessly and in a user-friendly design, which ultimately positively impacts the cost-time-effectiveness of their decision-making processes.

Conclusion

Smart cities go beyond the use of digital technologies for better resource utilization and urban development, with data serving as a foundational element driving progress into actionable knowledge. Public data presents both opportunities and hurdles: while data accessibility fuels innovation, challenges such as data quality, interoperability, standardization, and privacy demand a careful navigation. Thus, to overcome each challenge, a strategic approach is needed to transform obstacles into pathways for progress. A solution emerges in the form of a public portal built upon a robust data warehouse infrastructure, assimilating diverse public data sources. This practical approach aims to extract the maximum value from open datasets, revolutionizing information management and fostering transparency and strategic decision-making. A centralized repository with data spanning different cities, including health, education, and mobility, benefits citizens, investors seeking business insights, and students shaping their education. As smart cities evolve, this proposed solution guides stakeholders towards a sustainable, inclusive, and data-driven urban future.

Notes

1 CIT website available at https://www.inteligenciaterritorial.pt/, accessed in 2024-06-06.
2 INE website available at https://www.ine.pt/, accessed in 2024-06-06.
3 SPMS website available at https://www.spms.min-saude.pt/, accessed in 2024-06-06.
4 IPMA website available at https://www.ipma.pt/, accessed in 2024-06-06.
5 E-Redes website available at https://www.e-redes.pt/, accessed in 2024-06-06.
6 Turismo Portugal website available at https://www.turismodeportugal.pt/, accessed in 2024-06-06.
7 Dados.Gov website available at https://dados.gov.pt/, accessed in 2024-06-06.

References

Agrawal, S., Chaudhuri, S., & Narasayya, V. (2000). Automated selection of materialized views and indexes for SQL databases. *Proceedings of the 26th International Conference on Very Large Data Bases, VLDB'00.*

Araujo, V., Mitra, K., Saguna, S., & Åhlund, C. (2019). Performance evaluation of FIWARE: A cloud-based IoT platform for smart cities. *Journal of Parallel and Distributed Computing*, *132*, 1–12. https://doi.org/10.1016/j.jpdc.2018.12.010

Attard, J., Orlandi, F., Scerri, S., & Auer, S. (2015). A systematic review of open government data initiatives. *Government Information Quarterly*, *32*(4). https://doi.org/10.1016/j.giq.2015.07.006

Becker, L. T., & Gould, E. M. (2019). Microsoft power BI: Extending excel to manipulate, analyze, and visualize diverse data. *Serials Review*, *45*(3). https://doi.org/10.1080/00987913.2019.1644891

Bellatreche, L. (2012). Dimension table selection strategies to referential partition a fact table of relational data warehouses. *Recent Trends in Information Reuse and Integration*, *1*, 19–41. https://doi.org/10.1007/978-3-7091-0738-6_2

Bibri, S. E., & Krogstie, J. (2017). Smart sustainable cities of the future: An extensive interdisciplinary literature review. *Sustainable Cities and Society*, *31*. https://doi.org/10.1016/j.scs.2017.02.016

Bilal, M., Sheikh, U., Raza, B., & Javaid, Q. (2016). Application of data warehouse in real life: State-of-the-art survey from user preferences' perspective. *International Journal of Advanced Computer Science and Applications*, *7*(4). https://doi.org/10.14569/ijacsa.2016.070455

Box, A. (2022). Introducing charticulator for power BI: Design vibrant and customized visual representations of data. In *Introducing Charticulator for Power BI: Design vibrant and customized visual representations of data*. Berkeley, CA: Apress. https://doi.org/10.1007/978-1-4842-8076-8

Braunschweig, K., Eberius, J., Thiele, M., & Lehner, W. (2012). The state of open data limits of current open data platforms. *Proceedings of the 21st World Wide Web Conference 2012, Web Science Track at WWW'12*.

Capgemini Institute, & European Commission. (2015). *Creating value through open data: study on the impact of re-use of public data resources*. European Commission.

Currie, M. (2020). Data as performance – Showcasing cities through open data maps. *Big Data and Society*, *7*(1). https://doi.org/10.1177/2053951720907953

De Dutta, S., & Prasad, R. (2020). Digitalization of global cities and the smart grid. *Wireless Personal Communications*, *113*(3). https://doi.org/10.1007/s11277-020-07478-w

Gardner, S. R. (1998). Building the data warehouse. *Communications of the ACM*, *41*(9). https://doi.org/10.1145/285070.285080

Gharaibeh, A., Salahuddin, M. A., Hussini, S. J., Khreishah, A., Khalil, I., Guizani, M., & Al-Fuqaha, A. (2017). Smart cities: A survey on data management, security, and enabling technologies. *IEEE Communications Surveys and Tutorials*, *19*(4). https://doi.org/10.1109/COMST.2017.2736886

Guerra, J., & Andrews, D. (2011). *Why you need a data warehouse*. Andrews Consulting Group (ACG).

Jaakkola, H., Mäkinen, T., & Eteläaho, A. (2014). Open data - opportunities and challenges. *ACM International Conference Proceeding Series*, *883*. https://doi.org/10.1145/2659532.2659594

John Walker, S. (2014). Big data: A revolution that will transform how we live, work, and think. *International Journal of Advertising*, *33*(1). https://doi.org/10.2501/ija-33-1-181-183

Kimball, R., Caserta, J., Kimball, R., & Caserta, J. (2004). *The data warehouse ETL toolkit: Practical techniques for extracting, cleaning conforming, and delivering data*. Wiley.

Kimball, R., Ross, M., & Anisimov, A. A. (2003). The data warehouse toolkit: The complete guide to dimensional modeling (2nd edition). *SIGMOD Record*, *32*(3). https://doi.org/10.1145/945721.945741

Lai, C. S., Jia, Y., Dong, Z., Wang, D., Tao, Y., Lai, Q. H., Wong, R. T. K., Zobaa, A. F., Wu, R., & Lai, L. L. (2020). A review of technical standards for smart cities. *Clean Technologies*, *2*(3). https://doi.org/10.3390/cleantechnol2030019

Law, K. H., & Lynch, J. P. (2019). Smart city: Technologies and challenges. *IT Professional*, *21*(6). https://doi.org/10.1109/MITP.2019.2935405

Lousa, A., Pedrosa, I., & Bernardino, J. (2019). Evaluation and analysis of business intelligence data visualization tools. *Iberian Conference on Information Systems and Technologies, CISTI, 2019-June*. https://doi.org/10.23919/CISTI.2019.8760677

Midor, K., & Płaza, G. (2020). Moving to smart cities through the standard indicators ISO 37120. *Multidisciplinary Aspects of Production Engineering*, *3*(1). https://doi.org/10.2478/mape-2020-0052

Moustaka, V., Vakali, A., & Anthopoulos, L. G. (2019). A systematic review for smart city data analytics. *ACM Computing Surveys*, *51*(5). https://doi.org/10.1145/3239566

Nambiar, A., & Mundra, D. (2022). An overview of data warehouse and data lake in modern enterprise data management. *Big Data and Cognitive Computing*, *6*(4). https://doi.org/10.3390/bdcc6040132

Neves, F. T., de Castro Neto, M., & Aparicio, M. (2020). The impacts of open data initiatives on smart cities: A framework for evaluation and monitoring. *Cities*, *106*. https://doi.org/10.1016/j.cities.2020.102860

Olivares-Rojas, J. C., Reyes-Archundia, E., Gutierrez-Gnecchi, J. A., Molina-Moreno, I., Cerda-Jacobo, J., & Mendez-Patino, A. (2021). A comparative assessment of embedded databases for smart metering systems. *2021 IEEE PES Innovative Smart Grid Technologies Conference - Latin America, ISGT Latin America 2021*. https://doi.org/10.1109/ISGTLatinAmerica52371.2021.9543058

Pereira, G. V., Macadar, M. A., Luciano, E. M., & Testa, M. G. (2017). Delivering public value through open government data initiatives in a Smart City context. *Information Systems Frontiers*, *19*(2). https://doi.org/10.1007/s10796-016-9673-7

Provost & Fawcett. (2013). Data science for business what you need to know about data mining and data-analytic thinking. *Journal of Chemical Information and Modeling*, *355*, 1–17.

Przybyłowski, A., Kałaska, A., & Przybyłowski, P. (2022). Quest for a tool measuring urban quality of life: ISO 37120 Standard sustainable development indicators. *Energies*, *15*(8). https://doi.org/10.3390/en15082841

Rahul, K., & Banyal, R. K. (2020). Data life cycle management in big data analytics. *Procedia Computer Science*, *173*. https://doi.org/10.1016/j.procs.2020.06.042

Runtuwene, J. P. A., Tangkawarow, I. R. H. T., Manoppo, C. T. M., & Salaki, R. J. (2018). A comparative analysis of extract, transformation and loading (ETL) process. *IOP Conference Series: Materials Science and Engineering*, *306*(1). https://doi.org/10.1088/1757-899X/306/1/012066

Sarker, I. H. (2021). Data science and analytics: An overview from data-driven smart computing, decision-making and applications perspective. *SN Computer Science*, *2*(5). https://doi.org/10.1007/s42979-021-00765-8

Souibgui, M., Atigui, F., Zammali, S., Cherfi, S., & Yahia, S. Ben. (2019). Data quality in ETL process: A preliminary study. *Procedia Computer Science*, *159*. https://doi.org/10.1016/j.procs.2019.09.223

Tardieu, H. (2022). Role of Gaia-X in the European data space ecosystem. In *Designing data spaces*. Cham: Springer. https://doi.org/10.1007/978-3-030-93975-5_4

Verhulst, S., Noveck, B. S., Caplan, R., Brown, K., & Paz, C. (2018). The open data era in health and social care. *NYLS Legal Studies Research Paper*, (2563788), 18-41. https://doi.org/10.2139/ssrn.2563788

Verhulst, S. G., & Young, A. (2017). Open data in developing economies: Toward building an evidence base on what works and how. In *Open data in developing economies: Toward building an evidence base on what works and how*. https://doi.org/10.47622/9781928331599

Watson, H. J. (2002). Recent developments in data warehousing. *Communications of the Association for Information Systems*, *8*(1), 1–3. https://doi.org/10.17705/1cais.00801

Williamson, B. (2015). Educating the smart city: Schooling smart citizens through computational urbanism. *Big Data and Society*, *2*(2). https://doi.org/10.1177/2053951715617783

Wing, J. M. (2019). The data life cycle. *Harvard Data Science Review*, *1*(1), 1–4 https://doi.org/10.1162/99608f92.e26845b4

Zhang, J., Symons, J., Agapow, P., Teo, J. T., Paxton, C. A., Abdi, J., Mattie, H., Davie, C., Torres, A. Z., Folarin, A., Sood, H., Celi, L. A., Halamka, J., Eapen, S., & Budhdeo, S. (2022). Best practices in the real-world data life cycle. *PLOS Digital Health*, *1*(1). https://doi.org/10.1371/journal.pdig.0000003

6

STRENGTHENING GOVERNANCE ACCOUNTABILITY IN SMART CITIES

The case of urban energy transition in Jakarta, Indonesia

Elena Reshetova and Kris Hartley

Introduction

According to the United Nations (UN), energy plays a fundamental role in the world's major policy challenges and opportunities (Filho et al., 2015). This policy imperative is reflected in Goal 7 of the UN Sustainable Development Goals (SDGs): to 'ensure access to affordable, reliable, sustainable and modern energy for all' (United Nations, 2017). The needs and aspirations of the global community are dependent in part on the transition to cleaner energy systems, a systemic effort involving anticipatory solutions and stakeholder collaborations for addressing environmental, economic, political, and technical challenges at all scales.

Consuming roughly 75 percent of global primary energy (United Nations, 2017), cities are an increasingly relevant scale when considering global energy policy. City governments face a local political mandate to ensure that energy supply is affordable and reliable but are also under pressure from global institutions to address environmental issues and strengthen physical and social infrastructure for climate adaptation. Such efforts have a substantial energy component, particularly in innovation and market functionality (Droege, 2016). Given the substantial impact of urban activity on global energy and environmental systems, cities are taking centre stage as the loci of action for achieving energy transition and other sustainability objectives (Hartley, 2019).

The often-contradictory nature of modern policy imperatives (e.g., economic growth as against carbon reduction) requires a stable, coherent, and institutionalized accountability structure to address inefficiencies and transaction costs that can arise from malfeasance and malpractice

DOI: 10.4324/9781003439325-9

(Campante and Do, 2014). Examples are principal-agent failures, incomplete contracts, information asymmetries, regulatory capture, non-compliance, opportunism, and interpretive ambiguity (see Chrisidu-Budnik and Przedańska, 2017; Bajari and Tadelis, 2001). In viewing energy transition as a policy goal impacted by and dependent on private sector behaviour, the design of government accountability mechanisms should be informed by concepts like rule-of-law and institutional objectivity and functionality, while also retaining the flexibility to accommodate situational emergence and shifting political imperatives (Slaughter, 2017; Janssen and Van Der Voort, 2016).

This chapter explores urban energy transitions through the perspective of accountability among public and private sector actors. The multifaceted nature of urban energy transition beckons to all sectors, highlighting actor relationships as a relevant unit of analysis. In settings where property rights define relationships among actors, resources and efforts for establishing and protecting those rights can be classified as transaction costs (Barzel and Allen, 2023). This chapter examines how accountability varies among actors and how that variability impacts urban energy transitions. For this purpose, we define accountability as the legal, political, and/or social obligation of a party entrusted with specified responsibilities to fulfil those responsibilities or be interrogated and sanctioned if failing to meet them.

This chapter examines accountability through the case of Jakarta, Indonesia, which is representative of the types of energy policy challenges faced by megacities across the Global South. Like Southeast Asia's other major population centres (Manila, Bangkok, Kuala Lumpur, and Ho Chi Minh City), Jakarta is vulnerable to the impacts of climate change due to its coastal location, low elevation, stressed infrastructure, and concentration of disadvantaged people living in precarious conditions (see Firman et al., 2011). The situation does not appear likely to improve soon. According to *The New York Times* (2018), between 2018 and the end of the 21st century, Jakarta's number of days with a peak temperature above 32°C is expected to increase from 235 to 349. By studying governance accountability in a setting with such high urgency for energy transition, this chapter provides policy insights both for current policymaking and for anticipatory efforts in cities (e.g., the Global North) likely to face similar threats in the future. Accordingly, the chapter's research question is: what is the role of accountability in achieving sustainable urban energy transitions?

The chapter proceeds as follows. The literature review addresses current knowledge about urban energy transitions and accountability, followed by a description of the analytical framework, methodology, and background on the Jakarta case. A description and analysis of Jakarta's accountability mechanisms is then presented across three elements of urban energy transition. The chapter concludes with a summary of key findings and policy recommendations.

Literature review

Academic studies about urban energy systems have clustered around literature on urbanism, urban planning, engineering, and systems modelling. This body of research has generally adopted an understanding about urban energy systems that is based on the perspective of physical elements and the ways associated technologies operate (Keirstead, 2013). As a topic in the social sciences, including non-technical elements (e.g., social, political, and economic), urban energy systems did not attract much interest until the 2000s because the research agenda had previously been driven primarily by engineering perspectives. The interdisciplinary perspective emerged in part from the understanding of cities as settings where institutional, political, and policy activities occur (Brenner and Theodore, 2002; Logan and Molotch, 1987). Over time, discourses about the future of cities and urban infrastructure synthesized broader contextual and policy concerns like sustainability and smartness (Hartley and Im, 2022).

Although no consensus definition of urban energy systems has emerged, research has hybridized it with other concepts: urban sustainability (Bertone et al., 2007; Boone and Fragkias, 2013; Broto, 2017; MacKay, 2009; McGranahan and Satterthwaite, 2003), urban ecosystems and ecology (Facchini et al., 2017, 83–85; Grimm et al., 2008; Rohracher and Späth, 2013), urban metabolism (Barles, 2010; Brunner, 2007; Facchini et al., 2017; Goldstein et al., 2013; Kennedy et al., 2009; Newman, 1999; Pincetl et al., 2012), and low-carbon transitions and carbon management (Bulkeley et al., 2012; Dulal and Akbar, 2013; Echenique, 2017; Francesch-Huidobro, 2017; Gately and Hutyra, 2017; Gately et al., 2017; Kennedy et al., 2009; Lebel et al., 2007; Marcotullio et al., 2013; Newig and Moss, 2017; Shane and Graedel, 2000). In addition to these various perspectives, the concept of material flow analysis (MFA) can help define the boundaries of an urban energy system through demand-side and supply-side factors (see Grubler and Fisk, 2012; Leach et al., 1997; Ramaswami et al., 2012). This chapter takes the demand-side or consumption-based perspective, which accounts for energy resources exploited outside the city but consumed within it. That is, the consumption of energy and material goods in cities impacts geographic areas outside cities where manufacturing and resource extraction take place (Phdungsilp, 2006).

In accordance with this chapter's research question, two concepts beyond energy supply systems require consideration: energy security and environmental sustainability. These issues have been studied typically as national-scale rather than urban phenomena, but with the growing role of urban systems in energy transition, research has more recently begun to follow suit. Likewise gaining new visibility is the proliferation of practical policy applications of both concepts to cities. For example, in urban energy security research the following issues are receiving increased attention: (i) energy independence within urban system boundaries; (ii) transition from nationally coordinated to local, distributed, and autonomous infrastructure; and (iii) urban capacity to pursue an energy security agenda outlined by the global community despite constraints imposed by and conflicts with central government objectives. In contexts where the visibility of international and national climate action agendas is prominent (Mead, 2017), the objectives of energy security and environmental sustainability are increasingly coterminous at political and policy levels. Nevertheless, the two topics are rarely combined in analyses of urban energy systems – a gap that this study seeks to fill by analysing energy security and environmental issues as constituents of a broader energy transition agenda.

Within the context of urban energy governance, this study focuses on accountability. Accountability is said by scholarly and broader knowledge communities to play a crucial role in facilitating quality (or 'good') governance by resolving challenges associated with principal-agent failures, information asymmetries, regulatory capture, non-compliance, and opportunism. Broadly synonymous with the concept of responsibility, accountability with respect to governance and policymaking has been the subject of numerous studies. As argued by Mulgan (2000), Dowdle (2006), and Ebrahim and Weisband (2007), the eclectic mix of studies and relative ambiguity of the concept have worked against the development of a consensus understanding. According to Bovens (2007), the concept of accountability can be considered a 'dustbin' of good intensions and ambiguous ideas about governance.

The academic literature has typically analysed government accountability by scale and type. Regarding scale, accountability is typically classified as external (to the public) and internal (among actors within and across levels of government). Regarding type, Koppell (2005) specifies five dimensions of accountability: transparency, liability, controllability, responsibility, and responsiveness. Similarly, Zengerling (2018) identifies four pillars of accountability in the context of climate change mitigation: responsibility, transparency, assessment, and participation. Even with the variety of definitions and elements proposed, Bovens (2010) broadly categorizes these studies as normative, whereby accountability is seen in a positive light. Thus, accountability is often invoked in discussions about 'good' and 'sustainable' governance (Çalıyurt and Said, 2018; Crowther et al., 2015; Dhiman and Marques, 2018; Gordon, 2016) and is often treated methodologically as a dependent variable. Bovens (2010) further argues that other studies regard accountability as a practical mechanism and thus methodologically as an independent variable, with the focus on relations among accountors (those accountable) and accountees (see also Bovens, 2007; Pollitt, 2003). Within the broad literature on accountability is a subset of research focused on local government (Hong, 2017; Nyawo, 2017; Quinlivan et al., 2014; Yilmaz and Venugopal, 2013; Yilmaz and Guner, 2013). Generally, accountability in local governments of low- and middle-income countries is found to be lacking (Bardhan and Mookherjee, 2006; Shah, 2006). The literature on accountability in urban energy and environmental management is largely normative and therefore concerned with assessment and performance rather than accountability as a mechanism for other objectives (Mees and Driessen, 2019; Zengerling, 2018; Hendriks, 2008; Baud and Dhanalakshmi, 2007).

Finally, studies of urban energy systems in countries outside the Organization for Economic Cooperation and Development (OECD) are scarce, due in part to limited external research interest and a perception that the most advanced or innovative systems are found only in high-income countries. This scarcity illuminates a troublesome gap as cities like Jakarta, Bangkok, and Manila represent some of the most energy-stressed and environmentally vulnerable regions of the world. A review of literature about these three Southeast Asian cities identified two notable trends (see Appendix). First, most studies focus on a specific issue (e.g., waste management, transportation, and household electricity consumption) rather than on the broader socio-technical dimensions of urban energy systems. Second, most provide little or no deep analysis of policymaking and governance issues. In the modestly sized literature on energy transitions in these Southeast Asian cities, accountability is not analysed in a systematic way, whether in-depth or as a peripheral issue. There is a body of research on accountability in Southeast Asian countries at the national level, but urban energy governance remains largely overlooked (Van Belle and Mayhew, 2016; Yilmaz and Venugopal, 2013; Baud and Dhanalakshmi, 2007; Yilmaz et al., 2007; Shah, 2006; Blair, 2000).

Methodology

Analytical framework

This study's research question is: what is the role of accountability in achieving sustainable urban energy transitions? To answer this question, the study applies the following three concepts: (i) accountability as a transaction cost associated with the governance structure of urban energy transition, (ii) governance structure as a distribution of property rights providing legal incentives to govern relationships, and (iii) transaction costs as the costs of establishing and maintaining these property rights. The chapter adopts Bovens' (2007) framework for analysing accountability, which views accountability as a mechanism and set of independent variables. The framework consists of definitions of accountability relations and guiding questions.

The framework proposes that the establishment of accountability relations is dependent on three elements adapted from Bovens (2010; 2007): (i) obligatory or voluntary willingness to inform (i.e., an actor is obligated or willing to share data with a given forum), (ii) the ability to question (i.e., the forum can question the adequacy and legitimacy of an actor's conduct), and

(iii) conduct appraisal and sanctioning ability (i.e., the forum may impose explicit or implicit rewards or penalties based on an actor's conduct). Notable are terms qualifying the above elements (i.e., 'voluntary,' 'potentially,' 'implicit,' and 'may'). These terms imply the fluid nature of accountability processes, particularly how the forum may choose whether to exercise its prerogative to question and assess actor conduct. Opting in and out of entitlements reflects a given level of trust in actor-forum relationships: constant monitoring at one extreme and complete trust at the other (Davies, 2001). Once established, accountability relations can be classified by determining the actor (accountable as an accountor), forum (receiving the account as an accountee), content, and purpose of the relationship (Table 6.1).

While accountability is conceptualized in this study as an independent variable, governance of urban energy transitions is conceptualized as a dependent variable. Energy transition is generally preconditioned by a motivation to combine energy security with environmental protection objectives. Cities must balance energy supply stability and security with concerns about sustainability, particularly as awareness and political pressure build around environmental issues. In the context of this case study, urban energy transition refers to the replacement of more-polluting energy sources with less-polluting energy sources, as realized through local policies across the three

TABLE 6.1 Types of accountability relations

Element	*Element identification process*	*Type of accountability*
Actor	Who is providing the account?	Corporate Hierarchical Collective Individual
Forum	Who is receiving the account?	Political Legal Administrative Professional Social
Content	What is in the account?	Legal Professional Financial Administrative Procedural Product
Link	Why is there an accountability relation?	Vertical Diagonal Horizontal

Note: Accountability relations can be defined on each of the four elements individually.
Source: Adapted from Bovens (2007).

domains most closely related to intensity of energy use (identified through interviews and in the literature): (i) electricity grids, (ii) building efficiency, and (iii) transportation.

Methodology

The chapter examines Jakarta, Indonesia, as a case of urban energy transition dynamics. Jakarta is representative of many of the climate-related challenges faced by the world's most rapidly developing cities (see Hasnath, 2024; Baker, 2012; Fuchs, 2010). Outside of Southeast Asia, examples of cities facing similar challenges are Lagos, Dhaka, Chennai, Mumbai, Karachi, and Rio De Janeiro, among many others. Although the particulars of governance arrangements vary across cases, this study of Jakarta is valuable in producing theoretical insights about accountability that can be tailored to explain dynamics elsewhere.

The qualitative single-case-study approach is consistent with the inductive logic of this study, fits its objectives, and allows for flexible investigation of the research question. Case study methods can be applied in a variety of contexts, and researchers across the social sciences have found the approach useful for qualitative research that requires complex contextual and historical analysis (George and Bennett, 2005). Justifications for qualitative study include the opportunity to investigate under-studied or misunderstood phenomena, to generate hypotheses and research propositions based on in-depth data analysis, and to focus on discovering and analysing patterns rather than measuring variables and causal mechanisms alone (Morse and Mitcham, 2016; Bettis et al., 2015; George and Bennett, 2005). As such, the qualitative approach serves this study by facilitating a systematic understanding of accountability processes through structured analysis and descriptive accounts of policymaking processes. At the same time, there are trade-offs and limitations associated with the case study method, including parsimony versus richness, representativeness, ability to estimate and measure items of interest, and external validity (George and Bennett, 2005). Attempts are made, in this case application and analysis to eliminate or minimize these issues.

Regarding urban energy transitions as a nascent phenomenon, the advantages of case studies are the ability to conduct process-tracing, to identify new hypotheses, variables, and causal mechanisms (George and Bennett, 2005; Bennett, 2004; Van, 1997; Collier, 1993), and to address equifinality (Levy, 2008; George and Bennett, 2005; Bennett, 2004). This approach contributes to the specification of variables and intervening causal mechanisms (equifinality), and to the identification of reciprocal causation and endogeneity effects. Finally, given the limited city-level data and secondary research about Southeast Asian cities, including Jakarta, the case study methodology is deemed appropriate as it enables in-depth exploration of context. Data have been collected through policy documents, field visits, and interviews with practitioners, experts, and researchers.

Case background: Jakarta

Recognized as one of Southeast Asia's economic powerhouses, Indonesia has joined the group of trillion-dollar economies, with a GDP exceeding US$1 trillion in 2018 (Sipahutar, 2018; International Monetary Fund, 2017). Indonesia has numerous characteristics that enable steady economic growth, including an abundance of natural resources, large domestic consumer market, and evolving regulatory mechanisms that support investment (Kurniawan and Managi, 2018). However, the country also suffers from challenges concerning energy delivery. As an archipelagic nation with 17,000 islands and home to a growing population of over 275 million people, Indonesia struggles to broaden energy access (Bland, 2017; World Bank, 2016). The quantitative challenges of projecting energy supply-demand and import-export dynamics are complicated by qualitative dilemmas such as the high share of fossil fuels in the national energy mix (Simanjuntak et al., 2022), pressures of urbanization (Mardiansjah et al., 2021), environmental degradation (Prasetyanto and Sari, 2021), and the dual role of government in the energy sector as a provider (through state-owned enterprises) and regulator (see Husodo and Madalina, 2021; Wicaksono, 2008). As such, Indonesia faces the challenging task of balancing short- and long-term energy security, environmental sustainability, and social and economic objectives.

As the centre of national political and economic activity, Jakarta has avoided or minimized many energy access problems. Indonesia's capital city had by 2016 achieved a near-total electrification rate (99.99 percent; ADB, 2016) and enjoys relatively stable electricity supply. Blackouts are rare due to reserve capacity and result more from technical than supply problems. However, like other rapidly urbanizing mega-cities, Jakarta has struggled to implement measures curbing greenhouse gas (GHG) emissions and protecting the natural environment from energy-related stressors. In 2005, for example, Jakarta accounted for over 34 million tons of CO_2 emissions, with a large share coming from electricity production and consumption. The largest sources of emissions by sector are industry (30 percent), residential (25), transportation (19), and commercial (15). The Jakarta Regional Development Planning Board has projected that, by 2030, CO_2 emissions under a business-as-usual scenario will be triple those in 2005 (Prima Diyarni, 2018).

Findings

This study's findings are classified based on the three elements of urban energy transition previously mentioned (electricity grids, building efficiency, and transportation). Policy implications are likewise ordered.

Electricity grids

Management of Jakarta's energy field at the local level is complex. Administratively, Jakarta is a special-status province as the capital city. Under the Regional People's Representative Council (DKI Jakarta), the city's legislative body, there is no institutionalization of responsibility for energy mirroring that of the Ministry of Energy and Mineral Resources (MEMR). There is also no energy department within the governor's office. At the provincial level, there exists an equivalent of the General Plan of National Energy (RUEN), called the General Plan of Local Energy (RUED) (Wafa, 2017). RUEN encompasses multiple agencies in addition to MEMR and provides guidelines for long-term planning in national energy development. In accordance with the current RUEN, last updated in 2021 (IESR, 2020), provincial governments must prepare and submit their respective RUEDs to the National Energy Council (DEN), an independent consultative government body.

A RUED cannot contradict RUEN targets and objectives (Sundaryani, 2017). In accordance with the RUEN, by 2025 the combined share of fossil fuels in the energy mix was mandated to be reduced by one quarter. The central government has proposed a number of initiatives to promote renewable energy, including lower fossil fuel subsidies, rural electrification programmes that incorporate renewable energy, and a solar feed-in-tariff policy (see Santika et al., 2020; Pardede, 2018). Further actions include the country's largest solar power plant (5MW), completed in 2015 in Kupang (Kosasih, 2016); the largest solar power plant (2MW) in Sulawesi, completed in 2016 (Suhari, 2016); and the country's first wind farm (75WM), completed in 2018 (see Jensen and Asmarini, 2018).

At the same time, there remain challenges in deepening renewable energy commitment (Langer et al., 2021). First, the RUEN's renewable energy target faces complications related to existing electricity market design (Heffron et al., 2022) and contradictory regulations and electricity pricing systems (Kennedy, 2018) – factors that can disadvantage the economic feasibility of alternative electricity sources. Second, due to geographic constraints, high population density, and limited land availability, Jakarta's renewable energy options are limited largely to solar power. Indeed, the largest proportion of the city's electricity originates from coal- and gas-fired power plants that are located outside the city limits. Third, integration of renewables into the electricity grid raises additional challenges for the energy system, including the need for alterations to transmission, resource planning, and portfolio optimization. Fourth, there are concerns that renewable energy is made to compete with traditional fossil fuels and environmental externalities are not considered when the price is calculated. Finally, the role of Indonesia's State Electricity Company (PLN), the country's state-owned electricity distributor, in integrating renewables is conflicted, as it is a fuel supplier for diesel generators and an owner-operator of the majority of the country's legacy-technology power plants (Bridle et al., 2018).

The Jakarta government, as with all of Indonesia's local governments, does not have substantial influence over electricity policy and planning, as it is obligated to follow the requirements of the RUEN. Nevertheless, since the city is responsible for implementing national regulations and policies, Jakarta's government can propose new measures provided they align with national energy policy objectives – including those related to renewable energy. Furthermore, some regulations of DKI Jakarta supersede national regulations concerning low-carbon and energy-efficient activities. In 2009, Jakarta committed to a 30 percent reduction in GHG emissions by 2030 (Prima Diyarni, 2018). Governor Regulation #131, issued in 2012, proposed a regional action plan for achieving this commitment, while Governor Regulation #38, issued the same year, proposed measures for energy and building efficiency (DKI Jakarta, 2012). Notably, a series of regulations adopted from 2016 to 2018 (#19/2016, #50/2017, and #49/2018; see Hamdi, 2019) underscore the monopolistic dominance of PLN – a factor seen to inhibit the emergence of small-scale energy producers that might offer alternative sources.

Building efficiency

In 2017, only 16 percent of Indonesia's final energy use was covered by mandatory energy efficiency policies (IEA, 2020). For example, under Government Regulation #70/2009 on energy conservation, much labelling for appliance efficiency is voluntary (including for air conditioners, refrigerators, and freezers) (IEA, 2019). Given that transport and residential buildings have the lowest regulatory coverage among energy usage types, Jakarta's energy efficiency fails to meet standards that would advance sustainability transition goals.

Unlike electricity supply, energy efficiency can be regulated at the local level provided the standards equal or exceed national requirements. The aforementioned Government Regulation #38/2012 is one of several instruments targeting energy efficiency in Jakarta's buildings. Understood to be a regulation for 'green' buildings, it addresses site planning, construction, utilization, maintenance, and deconstruction (DKI Jakarta, 2012). As with many regulations, however, lack of enforcement capacity compromises the effectiveness of this regulatory effort, and in 2019 the regulation was placed under review. In the absence of government capacity, some developers have used the label 'green building' without full proof of compliance. At the same time, the Green Building Council Indonesia, an NGO (non-governmental organization) established in 2009, has been filling the gap by certifying buildings as 'environmentally friendly' on its own terms. DKI Jakarta has largely ignored the 'green' criterion when issuing building permits but stated plans to enforce standards after a regulatory revision.

Adding institutional emphasis on Jakarta's energy conservation is Grand Design of Green Building Jakarta ('Grand Design'), an initiative that aims to ameliorate functional fragmentation among city departments and improve commitment and consensus among stakeholders (Mungkasa, 2019). Neither technical guidance nor a master plan, Grand Design is a roadmap for the formulation of government plans and strategies. Unlike conventional policies, it focuses first on the identification of a policy problem rather than on the relevant policymaking body. Its goal is to reduce CO_2 emissions and the consumption of energy and water in buildings by 30 percent by 2030 (Jakarta Green Building, 2020). Supplementing efforts by DKI Jakarta on building efficiency are the efforts of national-level ministries. The Ministry of Public Works and Housing develops city plans and oversees civil engineering works, including building construction, in all urban areas of Indonesia (see Permatasari et al., 2020).

Transportation

According to 2016 data (Ginting, n.d.), 98.9 percent of Jakarta's 9.9 million motor vehicles are private. A recent study by Sefriyadi et al. (2023) indicates that private vehicle ownership in Indonesia continues to rise, with Jakarta having the highest level. While the city's first public bus system, Transjakarta, has surpassed 1 million daily users (Syakriah, 2023), its 13 corridors have limited geographic reach (Christianto, 2019) – an increasing problem as the city continues to sprawl outwards to accommodate population growth. The city's first MRT line opened in 2019, with additional lines under construction or planned. Trams are slated to replace or supplement Transjakarta buses in another effort to boost capacity, broaden coverage, and diversify the transport mode profile (News Desk, 2018). Given that private vehicles will likely remain Jakarta's dominant mode in the near future, traffic control measures have also been introduced, including an even-odd plate number policy (News Desk, 2018) that in 2018 replaced the 3-in-1 high-occupancy vehicle lane programme begun in 2003 (The Jakarta Post, 2018). Designed as a transitional measure, the policy was anticipated to be replaced by an electronic road pricing scheme (ERP) similar to that operating in Singapore. The ERP project entered the tendering stage in 2019 (PricewaterhouseCoopers, 2019) but faced significant debates about political responsibility, legal challenges (The Jakarta Post, 2020b), uncertainty about time-frame, and implementation feasibility.

Data from Bappenas (Ministry of National Development Planning) indicate that roughly Rp 100 trillion ($6.1 billion) is wasted each year as a consequence of traffic congestion and lagging transport infrastructure capacity (Ganesha, 2017). Jakarta's transportation sector is also a significant source of emissions, contributing roughly 20 percent of overall CO_2 emissions; this share is expected to grow to 25 percent by 2030 under a business-as-usual scenario (Prima Diyarni, 2018). The Indonesian and DKI Jakarta governments have acknowledged that this problem results from high growth in the number of vehicles, low road capacity-to-vehicle ratio, and low availability and use of public transport (Prima Diyarni, 2018). Hence, three transportation policy domains requiring further policy attention in the context of energy transition are public transport (e.g., metro/heavy rail, light rail, and bus rapid transit), traffic management (e.g., congestion pricing and restrictions on vehicle use), and road capacity (e.g., network development, pedestrian infrastructure, and technology-based initiatives).

The broader context of infrastructure policy in Jakarta is shaped by multiple layers of government administration. The city of Jakarta is part of the broader metropolitan region of Jabodetabek, which has been governed by the Jabodetabek Development Cooperation Agency (BKSP) since 1975. Jointly headed by the governors of Jakarta, West Java, and Banten, it also shares power with the heads of district and municipal governments across the region. While it has broad scope, BKSP lacks substantial enforcement power, has little authority over local governments, and is essentially ineffective (Firman, 2014). DKI Jakarta has funded environmental initiatives, including waste management and river clean-up, in various regions of Jabodetabek since 2010. This has included US$29 million per year to Bekasi for landfill management, US$520,000 per year to Bogor Regency for river-cleaning and catchment maintenance, US$2.7 million per year to Depok City for lake-cleaning, and US$1.75 million per year to Tangerang Regency for river-cleaning (Mungkasa, 2019).

Jakarta's transportation planning falls under the mandate of the Greater Jakarta Transportation Agency (BPTJ), an administrative body within the Ministry of Transportation having authority over transit activities in the Jakarta area (Ganesha, 2017). Under BPTJ, Jakarta city government has been engaged in various resource-sharing programmes. For example, in 2015 Jakarta committed US$4.5 million for revitalizing a BRT terminal in South Tangerang City. Jakarta also subsidizes BRT operational costs for routes serving surrounding municipalities. In 2019, the formation of a 'Greater Jakarta transportation authority' was announced (The Jakarta Post, 2019), but as of 2024 it appears that the authority is not fully operational.

Fossil fuels account for a majority of energy sources for Jakarta's transportation sector. However, efforts to promote biofuels have been undertaken (McDonald and Rahmanulloh, 2018), and electric vehicles have emerged as an energy transition pathway. As part of Presidential Regulation #22/2017, a national target was set for 2.1 million electric motorbikes and 2,200 hybrid or electric cars by 2025 (Reinard Sulaiman, 2019). Further, Transjakarta began trialing electric buses in 2019 and placed two into operation in early 2020 (The Jakarta Post, 2020a; Atika, 2019). In 2019, the city's largest taxi service, Blue Bird, introduced 30 electric taxis (e-taxis) built by BYD and Tesla (Nyoman Wira, 2019). Despite these efforts, infrastructure challenges accompanying transport electrification are substantial, particularly regarding the development of a threshold supply of charging stations in optimal locations.

Combined insights

This investigation of Jakarta's electricity grid, building efficiency, and transportation sector identifies commonalities, interdependence, and overlap among governance and administrative structures, resource and operational constraints, and numerous and sometimes contradictory stakeholder interests. The variety of actors within each energy policy domain suggests the presence of numerous accountability relations, a situation that can complicate energy transition. Notable also is the absence of explicit accountability relations in some domains, including between energy consumers and central government. Most other identified accountability relations (except two related to transportation) are vertical and based on either principal-agent relationships or legal and regulatory obligations.

The most clearly defined and diverse accountability relations exist in Jakarta's transportation sector. Only here does the public (i.e., transportation users) represent the forum as a clear accountee. The absence of the public as an accountee in the two other sectors (electricity grid and building efficiency) potentially hampers the pursuit of energy transition through democratic or participatory mechanisms. An informed public can provide a crucial check on the behaviours of the state, particularly in sectors where large-scale private enterprises dominate markets and have strong influence on policymaking. For electricity, the accountability link of PLN to the public is not direct because it is mediated through PLN's accountability to central government. This arrangement shifts the burden of accountability from local to national, engendering a mismatch between the scale of service delivery and that of feedback mechanisms.

This analysis demonstrates how accountability relations impact urban energy transitions. Based on progress towards energy transition, lagging sectors (in this case, electricity grid and building efficiency) have the weakest or most obscure accountability relationships; transportation, which shows the most progress among the three towards energy transition, has the most. At the same time, local government accountability does not alone contribute to sustainability transition. The cumulative effect of numerous relationships also plays a crucial role, across what Bovens (2007) conceptualizes as vertical, diagonal, and horizontal links. In Jakarta, the multilayered structure of urban service provision through local and regional bodies and national ministries creates a web of accountability links that serve as a check on the dominance of any single actor. However, the links found in this study are often thin from the perspective of the public.

Conclusions and policy implications

The world stands at a crossroads with regard to climate change and the pursuit of ecological health and stability. Efforts like the UN SDGs have focused global policy efforts and provide a template for the nationalization and localization of global sustainability visions. With global crises now at the doorstep of many cities, the urgency for urban policymakers to quickly and effectively address issues like energy transition is growing. No longer is it enough to deliver urban public services efficiently and effectively; city governments must now consider the longer-term impacts of urban systems on global climate conditions, while individuals must critically reckon with their own lifestyle and consumption choices.

This chapter has examined urban energy transition through the lens of governance accountability in Jakarta. This combination of topic and context is underexplored realm of the literature, and the case provides an instructive example of the complexities of actor relationships while typifying the multilayered governance settings that characterize many of the world's mega-cities. The case also combines a setting of high urgency for energy transition (given the city's vulnerability to climate change) with a governance paradigm rooted in the legacy of developmentalism and centralization (a circumstance common across Asia). While Indonesia has undertaken decentralization measures, there remains a web of accountability networks connecting local and national policy spheres, with legacy relationships and institutions continuing to shape (and sometimes constrain) policy action. This study finds that links with the public as a forum, however, are relatively scarce – potentially compromising the effectiveness, thoroughness, and legitimacy of energy transition while fossil fuel interests remain a dominant part of the national energy policy agenda.

The findings reveal several areas for further research that have the potential to positively impact practice. First, a cross-case comparison is needed to place the findings into broader contextual relief, including whether the accountability network landscape of Jakarta is similar to that of peer cities and whether differences in this landscape explain energy transition outcomes. Second, a deeper case analysis is needed to establish a stronger empirical basis for understanding accountability mechanisms within specific governance units. Such a study can include in-depth qualitative explorations of, for example, actor-forum relationships in public transit provision. Finally, future studies should seek to link Jakarta's manifold accountability mechanisms with broader theoretical questions about the institutionalized replication of power structures, focusing on how policies targeting energy transition are potentially rooted in status-quo interests, regressive social structures, and legacy policy epistemics that privilege technocratic ways of defining and solving problems. From an action-oriented perspective, such work can support push-back against what scholarship and practice often assume to be a benevolent and inclusive sustainability agenda – one that may actually be a reproduction of the same policy paradigms that originated and now accelerate the climate crisis.

This chapter concludes with three broad policy implications concerning the establishment and strengthening of accountability. These implications relate primarily to urban and local governments, while recognizing that national-level policies also impact urban energy transition. First, government units and agencies whose operations relate directly or indirectly to energy transition should commit to routine and institutionalized information-sharing. As 'actors' in the sense described by Bovens (2007), they are accountable to the public forum and therefore must ensure the transparency that clarifies priorities, procedures, and institutional relationships. The United Nations Human Settlement Programme (UN-HABITAT, 2004) provides a toolkit for improving transparency in urban governance, focused on five areas: assessment and monitoring for measuring improvement, access to information, ethics and integrity, institutional reforms that streamline procedures and strengthen participatory mechanisms, and a focus on specific issues or policy domains (e.g., energy transition) as examples and test-beds.

Second, efforts to improve transparency should enable the forum to freely question the adequacy and legitimacy of an actor's conduct. In the context of urban energy transition, this ability is dependent in part on democratic mechanisms and freedom of speech and press. Civil society must be empowered to observe government conduct and raise questions about its actions. The implication is that legitimacy emerges not from authoritarian mechanisms to limit criticism but from the ability of the public to challenge deficiencies in government processes and outcomes with the goal of fostering improvement.

Finally, there must be a mechanism to translate the collective sentiments of the forum into sanctioning actions. In a competitive or 'market-based' system, forum sanctioning might entail withholding of patronage to the offending actor; in colloquial terms, consumers 'speak with their dollar' by taking their business elsewhere. However, in most sectors related to energy transition, monopolies limit consumer choice and market entry by competitors. Such structures can result in monopolistic behaviours that lead to high prices and poor service; they also forestall competitive mechanisms as a means to strengthen commercial and political accountability. If such circumstances cannot be changed, non-market mechanisms must be developed to sanction actors believed by the public to engage in poor management (e.g., lagging efforts to pursue energy transition). When more accountability links between government actors and the public are established, sanctioning can have stronger effects by removing underperforming parties from power or sending signals that policy change is needed.

The description of these three policy implications implicitly suggests that each is linked to the other in a progressive sequence. Information and transparency enable the forum to credibly challenge an underperforming actor, which then justifies the possibility of punishment or sanctioning. No single one of these steps is effective without the other two, and the holistic nature of this policy approach is both its strength and its challenge. Indeed, making the public sector more transparent and democratically accountable is a topic that has occupied academic scholarship for decades and inspired proposals for public sector reform paradigms like New Public Administration (Frederickson, 1980) and New Public Service (Denhardt and Denhardt, 2000). While comprehensive reform may be unrealistic in complex governance settings with deep histories of centralized control, good-faith progress towards these and other approaches can begin to strengthen the building blocks of accountability.

Appendix:

TABLE 6.2 Survey of recent literature relevant to energy governance in Bangkok, Jakarta, and Manila

Publication	*Research subject*	*Methodology*
BANGKOK		
Phdungsilp (2006)	Sustainable development in Asian mega-cities	3-stage approach: Data review, LEAP for estimating energy demand, and MCDM model
Shrestha et al. (2008)	Energy consumption by slum dwellers in BKK	Literature review and household survey
World Bank (2009)	Climate change impact on BMR	Assessment and modelling of climate change consequences
Phdungsilp (2010)	Energy and carbon development pathways in BKK	Modelling (LEAP)
Prajongsan and Sharples (2012)	Building efficiency in BKK	CFD model
Lee and Van De Meene (2013)	Relationship between energy consumption and CO_2 emissions in cities, including BKK	Statistical modelling
Menikpura et al. (2014)	Co-benefits of landfill-gas-to-energy projects in BMA	Life cycle assessment and alternative scenario analysis
Berquist et al. (2015)	The role of individuals in climate change adaptation policies	Qualitative (secondary literature and interviews)

(*Continued*)

TABLE 6.2 (Continued)

Publication	*Research subject*	*Methodology*
Arifwidodo and Chandrasiri (2015)	Urban Heat Island effect and household energy consumption in BKK	Survey questionnaire
OECD (2015)	Policies to promote green growth in BMR	Literature and data review
Terry et al. (2016)	Coastal vulnerability of BKK	Coastal surveys and CCBs analysis
Thitanuwat et al. (2017)	The effect of urban green areas in BKK	Case study
Wongsapai et al. (2016)	Actions to achieve energy efficiency and renewable energy development in BMR	Scenario analysis
Ali et al. (2017)	CO_2 emissions in BMA	Modelling (MCDA)
Cheewaphongphan et al. (2017)	Transportation sector emissions in BKK	Modelling (GAINS)
Young (2016)	Green city policies in BKK	Literature review
Jareemit and Limmeechokchai (2017)	BKK residents' behaviour towards energy efficiency	Survey questionnaire
Singtong and Pharino (2018)	Waste management in BKK	Waste map development
Zhang and Li (2017)	Low carbon pathways for BKK and other Asian cities	Literature review
Eskew et al. (2018)	Rooftop solar PV system integration into the energy system of a BKK university	Life cycle assessment and scenario analysis
JAKARTA		
Hanson (1989)	Sustainability of urban areas, including JKT	Simulations
Steinberg (2007)	Current environmental sustainability developments in JKT	Literature review
Utama and Gheewala (2009)	Building efficiency and materials in JKT	Process analysis
Sovacool and Brown (2010)	Carbon footprints of JKT and other cities (including Manila)	4-step analysis of metropolitan areas and attributable carbon footprint
Kusno (2011)	Mobilization of various levels of government to produce a green environment in JKT	Discourse analysis
Turner (2012)	Gender and urban transportation in JKT	Surveys and interviews

(*Continued*)

TABLE 6.2 (Continued)

Publication	*Research subject*	*Methodology*
Hasibuan et al. (2014)	Transit Oriented Development (TOD) in JKT and Jabodetabek	Land use map and interview survey
Surahman et al. (2017)	Building sector in JKT	Case study and material flow analysis
Dewi et al. (2016)	JKT GHG emissions and reduction targets for 2030	Modelling (ExSS)
Rukmana (2018)	Rapid urbanization and sustainable transportation policies in JKT	Literature review
MANILA		
Gavieta (1997)	Socio-economic and spatial indicators of urban household energy consumption	Survey
Kelly (2003)	Policy choices of land conversion in MLA	Interviews and literature review
Porio and Crisol (2004)	Impact of property rights and tenure on poor urban households in MLA	Interviews, household survey, focus group discussions, and secondary data
Biona et al. (2007)	Carbon emissions of the two-stroke-powered tricycle in MLA	Instantaneous emissions model
Tiangco et al. (2008)	Urban Heat Island effect in MLA	Modelling (ASTER)
Serrona and Yu (2009)	Combination of community-based recycling and landfill management strategies for environmental sustainability in MLA	Evaluation, focus group discussion
Guzman (2010)	Quantifying sustainability in MLA	Indicator analysis
Porio (2011)	Resilience of urban poor households in MLA	Survey
Sahakian (2011)	Meeting growing energy demand in MLA	Data analysis
Sahakian and Steinberger (2011)	Energy consumption patterns and drivers at the household level in MLA	Semi-structured interviews
Sahakian (2012)	The role of the public in green economics transition in MLA	Semi-structures interviews and observations analysis
Ahanchian and Manuel Biona (2014)	Long-term benefits of reduced energy use in MLA	Modelling (LEAP)

(*Continued*)

TABLE 6.2 (Continued)

Publication	*Research subject*	*Methodology*
Porio (2014)	Urban poor households and the intersection of ecological-environmental and social vulnerabilities	Household surveys, interviews, focus group discussions, and secondary literature
Francisco (2015)	Benefits of improved air quality in MLA	CVM and survey
Mouton (2015)	Impact of neoliberal reforms on the energy sector in MLA	Case studies of two urban communities: interviews and primary data analysis
Edelman (2016)	Urban environmental management in the Philippines	Data and literature review
Gruyter et al. (2016)	Sustainable public transport in 26 cities, including Manila	Quantitative measurement framework
Andong and Sajor (2017)	Public transportation system as a source of emissions in MLA	Origin – destination matrices and qualitative data analysis
Estoque et al. (2017)	Relationship between land surface temperature and green space abundance in MLA, BKK, and JKT	Geospatial modelling (Landsat-8 OLI/TIRS and others)
Meerow (2017)	Climate change adaptation of coastal cities like MLA	Interviews, workshops
Esguerra (2018)	Sustainability in MLA	Data and literature review

Notes: For full titles see Bibliography.

Abbreviations: ASTER – Advanced Spaceborne Thermal Emission and Reflection Radiometer, BKK – Bangkok, BMA – Bangkok Metropolitan Area, BMR – Bangkok Metropolitan Region, CCB – coastal carbonate boulder, CFD – Computational fluid dynamics, CVM – Contingent valuation method, ExSS – Extended Snapshot tool, GAINS – Greenhouse Gas and Air Pollution Interactions and Synergies, GHG – Greenhouse has, JKT – Jakarta, LEAP – Long-range Energy Alternative Planning, MCDM – Multi-Criteria Decision-Making, MLA – Manila, OLI/TIRS – Operational Land Imager/Thermal Infrared Sensor, PV – photovoltaic.

Bibliography

"How Much Hotter Is Your Hometown than When You Were Born?" *The New York Times*, August 30, 2018. https://www.nytimes.com/interactive/2018/08/30/climate/how-much-hotter-is-your-hometown.html?smid=fb-nytscience&smtyp=cur

"Jakarta Green Building." Accessed March 19, 2020. https://greenbuilding.jakarta.go.id/index-en.html

"The Future We Want." United Nations, 2012. https://sustainabledevelopment.un.org/content/documents/733FutureWeWant.pdf

ADB. *Achieving Universal Electricity Access in Indonesia*. Manila: Asian Development Bank, 2016.

Ahanchian, Mohammad, and J.B. Manuel Biona. "Energy Demand, Emissions Forecasts and Mitigation Strategies Modeled over a Medium-range Horizon: The Case of the Land Transportation Sector in Metro Manila." *Energy Policy* 66 (March 1, 2014): 615–629. https://doi.org/10.1016/j.enpol.2013.11.026

Ali, Ghaffar, Nathsuda Pumijumnong, and Shenghui Cui. "Decarbonization Action Plans Using Hybrid Modeling for a Low-carbon Society: The Case of Bangkok Metropolitan Area." *Journal of Cleaner Production* 168 (December 2017): 940–951. https://doi.org/10.1016/j.jclepro.2017.09.049

Andong, Rebeca Fontanilla, and Edsel Sajor. "Urban Sprawl, Public Transport, and Increasing CO2 Emissions: The Case of Metro Manila, Philippines." *Environment, Development and Sustainability: A Multidisciplinary Approach to the Theory and Practice of Sustainable Development* 19, no. 1 (2017): 99–123.

Arifwidodo, Sigit, and Orana Chandrasiri. "Urban Heat Island and Household Energy Consumption in Bangkok, Thailand." *Energy Procedia* 79 (November 1, 2015): 189–194. https://doi.org/10.1016/j.egypro.2015.11.461

Atika, Sausan. "Transjakarta Introduces Electric Buses." *The Jakarta Post*, April 30, 2019. https://www.thejakartapost.com/news/2019/04/30/transjakarta-introduces-electric-buses.html

Bajari, P., and S. Tadelis. "Incentives versus Transaction Costs: A Theory of Procurement Contracts." *Rand Journal of Economics* 32, no. 3 (2001): 387–407.

Baker, J. L. (Ed.). *Climate Change, Disaster Risk, and the Urban Poor: Cities Building Resilience for a Changing World*. Washington D.C.: World Bank Publications, 2012.

Bardhan, Pranab and Dilip Mookherjee. "Decentralization, Corruption and Government Accountability." In *International Handbook on the Economics of Corruption*, edited by Susan Rose-Ackerman. Cheltenham; Northampton, MA: Edward Elgar Publications, 2006.

Barles, Sabine. "Society, Energy and Materials: The Contribution of Urban Metabolism Studies to Sustainable Urban Development Issues." *Journal of Environmental Planning and Management* 53, no. 4 (June 1, 2010): 439–455. https://doi.org/10.1080/09640561003703772

Barzel, Y., and D. W. Allen. *Economic Analysis of Property Rights*. Cambridge University Press, 2023.

Baud, Isa, and R. Dhanalakshmi. "Governance in Urban Environmental Management: Comparing Accountability and Performance in Multi-Stakeholder Arrangements in South India." *Cities* 2, no. 24 (2007): 133–147. https://doi.org/10.1016/j.cities.2006.11.003

Bennett, Andrew. "Case Study Methods: Design, Use, and Comparative Advantages." In *Models, Numbers, and Cases: Methods for Studying International Relations*. First edition., edited by Detlef F. Sprinz, and Yael Wolinsky-Nahmias. Ann Arbor: University of Michigan Press, 2004.

Berquist, Michelle, Amrita Daniere, and Lisa Drummond. "Planning for Global Environmental Change in Bangkok's Informal Settlements." *Journal of Environmental Planning and Management* 58, no. 10 (October 3, 2015): 1711–1730. https://doi.org/10.1080/09640568.2014.945995

Bertone, Genevieve, Parry, Shannon C., Kubani, D. and Jennifer Wolch. "Indicators in Action: The Use of Sustainability Indicators in the City of Santa Monica."

In *Community Quality-of-Life Indicators: Best Cases II*, edited by M. Joseph Sirgy, Don Rahtz, and David Swain. Springer Science & Business Media, 2007.
Bettis, Richard A., Alfonso Gambardella, Constance Helfat, and Will Mitchell. "Qualitative Empirical Research in Strategic Management: Editorial." *Strategic Management Journal* 36, no. 5 (May 2015): 637–639. https://doi.org/10.1002/smj.2317
Biona, J.B. Manuel, Alvin Culaba, and M.R.I. Purvis. "Energy Use and Emissions of Two Stroke-Powered Tricycles in Metro Manila." *Transportation Research Part D: Transport and Environment* 12 (October 1, 2007): 488–497. https://doi.org/10.1016/j.trd.2007.06.005
Blair, Harry. "Participation and Accountability at the Periphery: Democratic Local Governance in Six Countries." *World Development* 28, no. 1 (2000): 21–39.
Bland, Ben. "Indonesia Starts Count to Solve the Riddle of the Islands," *Financial Times*, June 3, 2017. https://www.ft.com/content/3acc43f0-45f5-11e7-8519-9f94ee97d996
Boone, Christopher G., and Michail Fragkias, eds. *Urbanization and Sustainability: Linking Urban Ecology, Environmental Justice and Global Environmental Change*. Human-Environment Interactions. Netherlands: Springer, 2013. https://doi.org/10.1007/978-94-007-5666-3
Bovens, Mark. "Analysing and Assessing Accountability: A Conceptual Framework1." *European Law Journal* 13, no. 4 (2007): 447–468. https://doi.org/10.1111/j.1468-0386.2007.00378.x
Bovens, Mark. "Two Concepts of Accountability: Accountability as a Virtue and as a Mechanism." *West European Politics* 33, no. 5 (September 1, 2010): 946–967. https://doi.org/10.1080/01402382.2010.486119
Brenner, Neil, and Nik Theodore. "Preface: From the 'New Localism' to the Spaces of Neoliberalism." *Antipode* 34, no. 3 (2002): 341–347. https://doi.org/10.1111/1467-8330.00245
Bridle, Richard, Philip Gass, Aidy Halimajaya, Lucky Lontoh, Neil McCulloch, Erica Petrofsky, and Lourdes Sanchez. *Missing the 23 Per Cent Target: Roadblocks to the Development of Renewable Energy in Indonesia*. International Institute for Sustainable Development, 2018. https://www.iisd.org/sites/default/files/publications/roadblocks-indonesia-renewable-energy.pdf
Broto, Vanesa Castán. "Energy Landscapes and Urban Trajectories towards Sustainability." *Energy Policy* 108 (September 1, 2017): 755–764. https://doi.org/10.1016/j.enpol.2017.01.009
Brunner, Paul H. "Reshaping Urban Metabolism." *Journal of Industrial Ecology* 11, no. 2 (2007): 11–13. https://doi.org/10.1162/jie.2007.1293
Bulkeley, Harriet, Vanesa Castán Broto, Mike Hodson, and Simon Marvin. *Cities and Low Carbon Transitions*. 1st Edition. Routledge, 2012. https://www.routledge.com/Cities-and-Low-Carbon-Transitions-1st-Edition/Bulkeley-Castan-Broto-Hodson-Marvin/p/book/9780415586979
Çalıyurt, Kıymet Tunca, and Roshima Said, eds. *Sustainability and Social Responsibility of Accountability Reporting Systems: A Global Approach*. Accounting, Finance, Sustainability, Governance & Fraud: Theory and Application. Springer Singapore, 2018. https://doi.org/10.1007/978-981-10-3212-7
Campante, F. R., and Q. A. Do. "Isolated Capital Cities, Accountability, and Corruption: Evidence from US States." *American Economic Review* 104, no. 8 (2014): 2456–2481.

Cheewaphongphan, Penwadee, Agapol Junpen, Savitri Garivait, and Satoru Chatani. "Emission Inventory of On-Road Transport in Bangkok Metropolitan Region (BMR) Development during 2007 to 2015 Using the GAINS Model." *Atmosphere* 8, no. 9 (September 2017): 167. https://doi.org/10.3390/atmos8090167

Chrisidu-Budnik, A., and J. Przedańska. "The Agency Theory Approach to the Public Procurement System." *Wroclaw Review of Law, Administration & Economics* 7, no. 1 (2017): 154–165.

Christianto, Yesaya. "Tramways in Jakarta: An Impossible Plan?" *The Jakarta Post*, February 28, 2019. https://www.thejakartapost.com/academia/2019/02/28/tramways-in-jakarta-an-impossible-plan.html

Collier, David. "The Comparative Method." In *Political Science: The State of the Discipline II*, edited by Ada W. Finifter. Washington, DC: Amer Political Science Assn, 1993.

Crowther, David, Mourad Oubrich, Redouane Barzi, and Sara Abdaless, eds. *Governance, Accountability and Sustainable Development*. Newcastle upon Tyne: Cambridge Scholars Publishing, 2015. https://www.cambridgescholars.com/governance-accountability-and-sustainable-development

Davies, Anne. *Accountability: A Public Law Analysis of Government by Contract*. Oxford: Oxford University Press, 2001. https://www.oxfordscholarship.com/view/10.1093/acprof:oso/9780198299486.001.0001/acprof-9780198299486

Denhardt, Robert B., and Janet Vinzant Denhardt. "The New Public Service: Serving Rather than Steering." *Public Administration Review* 60, no. 6 (2000): 549–559. https://doi.org/10.1111/0033-3352.00117

Dewi, Retno, Ucok Siagian, Iwan Hendrawan, Rizaldi Boer, Lukytawati Anggraeni, and Toni Bakhtiar. "Low-Carbon City Scenarios for DKI Jakarta Towards 2030." In *Low Carbon Urban Infrastructure Investment in Asian Cities*, London: Palgrave Macmillan, 61–76, 2016. https://doi.org/10.1057/978-1-137-59676-5_4

Dhiman, Satinder, and Joan Marques, eds. *Handbook of Engaged Sustainability*. Cham: Springer International Publishing, 2018. https://www.springer.com/gp/book/9783319713113

DKI Jakarta. "Governor Regulation of DKI Jakarta Number 38 of 2012 on Green Building." *Indonesia Real Estate Law* (blog), 2012. http://www.indonesiarealestatelaw.com/summary-of-regulations/governor-regulation-of-dki-jakarta-number-38-of-2012-on-green-building/

Dowdle, Michael W. "Public Accountability: Conceptual, Historical, and Epistemic Mappings." In *Public Accountability: Designs, Dilemmas and Experiences*, edited by Michael W. Dowdle. Cambridge: Cambridge University Press, 2006.

Droege, Peter. "Renewable Energy and the City: Urban Life in an Age of Fossil Fuel Depletion and Climate Change." *Bulletin of Science, Technology & Society*, July 26, 2016. https://doi.org/10.1177/0270467602022002003

Dulal, Hari Bansha, and Sameer Akbar. "Greenhouse Gase Emission Reduction Options for Cities: Finding the "Coincidence of Agendas" between Local Priorities and Climate Change Mitigation Objectives." *Habitat International* 38 (January 2013): 100–105. https://doi.org/10.1016/j.habitatint.2012.05.001

Ebrahim, Alnoor, and Edward Weisband, eds. *Global Accountabilities: Participation, Pluralism, and Public Ethics*. 1 edition. Cambridge; New York: Cambridge University Press, 2007.

Echenique, Marcial. "Low Carbon Cities: London Explorations," 9. Paris, 2017. https://www.eprg.group.cam.ac.uk/wp-content/uploads/2017/07/M.-Echenique.pdf

Edelman, David J. "Managing the Urban Environment of Manila." *Advances in Applied Sociology* 06, no. 03 (2016): 101–133. https://doi.org/10.4236/aasoci.2016.63010

Esguerra, Darryl John. "Manila among World's Least Sustainable Cities." *Inquirer*, November 2, 2018. https://globalnation.inquirer.net/170755/manila-among-worlds-least-sustainable-cities-report

Eskew, John, M. Ratledge, Michael L. Wallace, Shabbir H. Gheewala, and Pattana Rakkwamsuk. "An Environmental Life Cycle Assessment of Rooftop Solar in Bangkok, Thailand," 2018. https://doi.org/10.1016/j.renene.2018.02.045

Estoque, Ronald C., Yuji Murayama, and Soe W. Myint. "Effects of Landscape Composition and Pattern on Land Surface Temperature: An Urban Heat Island Study in the Megacities of Southeast Asia." *Science of the Total Environment* 577 (January 2017): 349–359. https://doi.org/10.1016/j.scitotenv.2016.10.195

Facchini, Angelo, Chris Kennedy, Iain Stewart, and Renata Mele. "The Energy Metabolism of Megacities." *Applied Energy* 186 (January 15, 2017): 86–95. https://doi.org/10.1016/j.apenergy.2016.09.025

Filho, Walter Leal, Arnolds Úbelis, and Dina Bērziņa, eds. *Sustainable Development, Knowledge Society and Smart Future Manufacturing Technologies*. World Sustainability Series. Springer International Publishing, 2015. https://doi.org/10.1007/978-3-319-14883-0

Firman, Tommy. "The Dynamics of Jabodetabek Development: The Challenge of Urban Governance." In *Regional Dynamics in a Decentralized Indonesia*, edited by Hal Hill. Singapore: Institute of Southeast Asian Studies, 2014. https://bookshop.iseas.edu.sg/publication/1985

Firman, T., I. M. Surbakti, I. C. Idroes, and H. A. Simarmata. "Potential Climate-change Related Vulnerabilities in Jakarta: Challenges and Current Status." *Habitat International* 35, no. 2 (2011): 372–378.

Francesch-Huidobro, Maria. "Legacies of Geopolitical Insularity for Urban Energy Transitions Today." Regional Project Energy Security and Climate Change Asia-Pacific (RECAP), December 19, 2017. https://www.kas.de/en/web/recap/single-title/-/content/das-erbe-politischer-isolation-fuer-heutige-energiepolitische-transformationen

Francisco, Jamil Paolo S. "Willingness To Pay For Air Quality Improvements From Using Electric Jeepneys In Metro Manila." *The Singapore Economic Review* 60, no. 04 (September 2015): 1550073. https://doi.org/10.1142/S0217590815500733

Frederickson, H. George. *New Public Administration*. University, AL: University of Alabama Press, 1980.

Fuchs, R. J. "Cities at Risk: Asia's Coastal Cities in an Age of Climate Change." *University of Hawaii at Manoa* 2010. <https://scholarspace.manoa.hawaii.edu/handle/10125/17646>

Ganesha, Amal. "Jakarta Transportation Body Lays Out Strategies to Fix the Capital's $7b Traffic Problem." *Jakarta Globe*, December 5, 2017. https://jakartaglobe.id/business/jakarta-transportation-body-lays-comprehensive-plan-fix-capitals-7b-traffic-problem

Gately, C. K., and L. R. Hutyra. "Large Uncertainties in Urban-Scale Carbon Emissions." *Journal of Geophysical Research: Atmospheres* 122, no. 20 (2017): 11242–11260. https://doi.org/10.1002/2017JD027359

Gately, Conor K., Lucy R. Hutyra, Scott Peterson, and Ian Sue Wing. "Urban Emissions Hotspots: Quantifying Vehicle Congestion and Air Pollution Using

Mobile Phone GPS Data." *Environmental Pollution* 229 (October 2017): 496–504. https://doi.org/10.1016/j.envpol.2017.05.091

Gavieta, Rommel C. "Socio-Economic and Spatial Indicators for Household Energy for a Tropical Urban Community in Urban Manila." *Building Research & Information* 25, no. 2 (March 1, 1997): 124–128. https://doi.org/10.1080/096132197370534

George, Alexander L., and Andrew Bennett. *Case Studies and Theory Development in the Social Sciences*. Fourth Printing edition. Cambridge, MA: The MIT Press, 2005.

Ginting, Ferdinand. "Toward A Better Jakarta Transportation (Transportation Agency, DKI Jakarta Provincial Government)." n.d. https://www.unescap.org/sites/default/files/4.1%20%20Planning%20and%20development%20of%20urban%20transport%20systems%20in%20Jakarta.pdf

Goldstein, Benjamin, Morten Birkved, Maj-Britt Quitzau, and Michael Hauschild. "Quantification of Urban Metabolism through Coupling with the Life Cycle Assessment Framework: Concept Development and Case Study." *Environmental Research Letters* 8, no. 3 (July 2013): 035024. https://doi.org/10.1088/1748-9326/8/3/035024

Gordon, David J. "The Politics of Accountability in Networked Urban Climate Governance." *Global Environmental Politics* 16, no. 2 (2016): 82–100.

Grimm, Nancy B., Stanley H. Faeth, Nancy E. Golubiewski, Charles L. Redman, Jianguo Wu, Xuemei Bai, and John M. Briggs. "Global Change and the Ecology of Cities." *Science* 319, no. 5864 (February 8, 2008): 756–760. https://doi.org/10.1126/science.1150195

Grubler, Arnulf, and David Fisk, eds. *Energizing Sustainable Cities: Assessing Urban Energy*. 1 edition. London; New York: Routledge, 2012.

Guzman, Maria Aileen Leah G. "Quantifying the Current and Future Sustainability of Metro Manila, Philippines." *State University of New York, College of Environmental Science and Forestry*, Dissertation, 2010.

Hamdi, Elrika. "Indonesia's Solar Policies: Designed to Fail?" *Institute for Energy Economics and Financial Analysis*, February 2019. https://ieefa.org/wp-content/uploads/2019/02/Indonesias-Solar-Policies_February-2019.pdf

Hanson, Mark E. "The Motor Vehicle Dominated City as a Non-Sustainable Urban Form: Mexico City and Jakarta." *Computers, Environment and Urban Systems* 13, no. 2 (January 1, 1989): 95–108. https://doi.org/10.1016/0198-9715(89)90037-9

Hartley, Kris. "Global Goals, Global Cities: Achieving the SDGs through Collective Local Action." *The Chicago Council on Global Affairs* 2019. https://www.thechicagocouncil.org/sites/default/files/report_global-goals-global-cities_190923.pdf

Hartley, Kris and Im, Tobin. "The Narrative Hegemony of Smart Governance: Social Change Through a Critical Theoretical Perspective." In *The Palgrave Handbook of Global Social Change*, edited by R. Baikady, S. M. Sajid, J. Przeperski, V. Nadesan, M. R. Islam, and J. Gao. London: Palgrave Macmillan, 2022.

Hasibuan, Hayati Sari, Tresna P. Soemardi, Raldi Koestoer, and Setyo Moersidik. "The Role of Transit Oriented Development in Constructing Urban Environment Sustainability, the Case of Jabodetabek, Indonesia." *Procedia Environmental Sciences* 20 (January 1, 2014): 622–631. https://doi.org/10.1016/j.proenv.2014.03.075

Hasnath, S. A. "Megacities in Developing Asia: Challenges and Opportunities." *The Elgar Companion to the Asian Development Bank* (2024): 167–176.

Heffron, R. J., M. F. Körner, T. Sumarno, J. Wagner, M. Weibelzahl, and G. Fridgen. "How Different Electricity Pricing Systems Affect the Energy Trilemma: Assessing Indonesia's Electricity Market Transition." *Energy Economics* 107 (2022): 105663.

Hendriks, Carolyn M. "On Inclusion and Network Governance: The Democratic Disconnect of Dutch Energy Transitions." *Public Administration* 86, no. 4 (2008): 1009–1031. https://doi.org/10.1111/j.1467-9299.2008.00738.x

Hong, Sounman. "What Are the Areas of Competence for Central and Local Governments? Accountability Mechanisms in Multi-Level Governance." *Journal of Public Administration Research and Theory* 27, no. 1 (January 1, 2017): 120–134. https://doi.org/10.1093/jopart/muw048

Husodo, J. A., and M. Madalina. "The Legal Risk to Sustainable Role of State-Owned Enterprises Management in Indonesia." *IOP Conference Series: Earth and Environmental Science* 724, no. 1, (2021, April): 012090.

IEA. "Energy Efficiency Awareness Raising – Policies," October 17, 2019. https://www.iea.org/policies/2458-energy-efficiency-awarenes-raising

IEA. "Indonesia - Countries & Regions." Accessed March 19, 2020. https://www.iea.org/countries/indonesia

IESR (Institute for Essential Services Reform). National Energy General Plan (RUEN) – Existing Plan, Current Policies, and Energy Transition Scenario (2020), 2020. <https://iesr.or.id/download/national-energy-general-plan-ruen-2020>

International Monetary Fund. "World Economic Outlook Database," October 2017. https://www.imf.org/external/pubs/ft/weo/2017/02/weodata/index.aspx

Janssen, M., and H. Van Der Voort. "Adaptive Governance: Towards a Stable, Accountable and Responsive Government." *Government Information Quarterly* 33, no. 1 (2016): 1–5.

Jareemit, Daranee, and Bundit Limmeechokchai. "Understanding Resident's Perception of Energy Saving Habits in Households in Bangkok." *Energy Procedia* 138 (October 1, 2017): 247–252. https://doi.org/10.1016/j.egypro.2017.10.048

Jensen, Fergus, and Wilda Asmarini. "Less Headwinds for Renewables Now Says Indonesia Pioneer." *Reuters*, February 12, 2018. https://www.reuters.com/article/us-indonesia-renewables-windfarm-idUSKBN1FW0W4

Keirstead, James. "Conceptualizing Urban Energy Systems." In *Urban Energy Systems: An Integrated Approach*. 1 edition, edited by James Keirstead, and Shah Nilay. Abingdon, Oxon; New York: Routledge, 2013.

Kelly, Philip F. "Urbanization and the Politics of Land in the Manila Region." *The ANNALS of the American Academy of Political and Social Science* 590, no. 1 (November 1, 2003): 170–187. https://doi.org/10.1177/0002716203256729

Kennedy, S. F. "Indonesia's Energy Transition and its Contradictions: Emerging Geographies of Energy and Finance." *Energy Research & Social Science* 41 (2018): 230–237.

Kennedy, Christopher, Julia Steinberger, Barrie Gasson, Yvonne Hansen, Timothy Hillman, Miroslav Havránek, Diane Pataki, Aumnad Phdungsilp, Anu Ramaswami, and Gara Villalba Mendez. "Greenhouse Gas Emissions from Global Cities." *Environmental Science & Technology* 43, no. 19 (October 1, 2009): 7297–7302. https://doi.org/10.1021/es900213p

Koppell, Jonathan GS. "Pathologies of Accountability: ICANN and the Challenge of 'Multiple Accountabilities Disorder.'" *Public Administration Review* 65, no. 1 (2005): 94–108. https://doi.org/10.1111/j.1540-6210.2005.00434.x

Kosasih, Felix Utama. "Kupang Houses Indonesia's Largest Solar Power Plant." *Global Indonesian Voices*, January 7, 2016. http://www.globalindonesianvoices.com/24393/kupang-houses-indonesias-largest-solar-power-plant/

Kurniawan, R., and S. Managi. "Economic Growth and Sustainable Development in Indonesia: An Assessment." *Bulletin of Indonesian Economic Studies* 54, no. 3 (2018): 339–361.

Kusno, Abidin. "The Green Governmentality in an Indonesian Metropolis." *Singapore Journal of Tropical Geography* 32, no. 3 (2011): 314–331. https://doi.org/10.1111/j.1467-9493.2011.00440.x

Langer, J., J. Quist, and K. Blok. "Review of Renewable Energy Potentials in Indonesia and their Contribution to a 100% Renewable Electricity System." *Energies* 14, no. 21 (2021): 7033.

Leach, Matthew, Ausilio Bauen, and Nigel Lucas. "A Systems Approach to Materials Flow in Sustainable Cities: A Case Study of Paper." *Journal of Environmental Planning and Management* 40 (November 1, 1997): 705–724. https://doi.org/10.1080/09640569711868

Lebel, Louis, Rodel D Lasco, Antonio Contreras, A P Mitra, Chhemendra Sharma, Hoang Tri Nguyen, Giok Ling Ooi, and Agus Sari. "Integrating Carbon Management into the Development Strategies of Urbanizing Regions in Asia." *Journal of Industrial Ecology* 11, no. 2 (2007): 61–81.

Lee, Taedong, and Susan Van De Meene. "Comparative Studies of Urban Climate Co-Benefits in Asian Cities: An Analysis of Relationships between CO2 Emissions and Environmental Indicators." *Journal of Cleaner Production* 58 (November 1, 2013): 15–24. https://doi.org/10.1016/j.jclepro.2013.04.047

Levy, Jack S. "Case Studies: Types, Designs, and Logics of Inference:" *Conflict Management and Peace Science*, March 1, 2008. https://doi.org/10.1080/07388940701860318

Logan, John R., and Harvey Luskin Molotch. *Urban Fortunes: The Political Economy of Place*. Berkeley: University of California Press, 1987. https://nyuscholars.nyu.edu/en/publications/urban-fortunes-the-political-economy-of-place

MacKay, David JC. *Sustainable Energy - Without the Hot Air*. 1 edition. Cambridge: UIT Cambridge Ltd., 2009.

Marcotullio, Peter John, Andrea Sarzynski, Jochen Albrecht, Niels Schulz, and Jake Garcia. "The Geography of Global Urban Greenhouse Gas Emissions: An Exploratory Analysis." *Climatic Change* 121, no. 4 (December 1, 2013): 621–634. https://doi.org/10.1007/s10584-013-0977-z

Mardiansjah, F. H., P. Rahayu, and D. Rukmana. "New Patterns of Urbanization in Indonesia: Emergence of Non-statutory Towns and New Extended Urban Regions." *Environment and Urbanization ASIA* 12, no. 1 (2021): 11–26.

McDonald, Garrett, and Arif Rahmanulloh. "Biofuels Annual Indonesia Biofuels Annual 2018." USDA Foreign Agricultural Service. Jakarta, Indonesia: USDA, August 13, 2018. https://apps.fas.usda.gov/newgainapi/api/report/downloadreportbyfilename?filename=Biofuels%20Annual_Jakarta_Indonesia_8-13-2018.pdf

McGranahan, Gordon, and David Satterthwaite. "Urban Centers: An Assessment of Sustainability." *Annual Review of Environment and Resources* 28, no. 1 (2003): 243–274. https://doi.org/10.1146/annurev.energy.28.050302.105541

Mead, Leila. "UN, Mayors Shine Spotlight on Sustainable Urban Development." *SDG Knowledge Hub: A Project by IISD* (blog), October 26, 2017. http://sdg.iisd.org/news/un-mayors-shine-spotlight-on-sustainable-urban-development/

Meerow, Sara. "Double Exposure, Infrastructure Planning, and Urban Climate Resilience in Coastal Megacities: A Case Study of Manila." *Environment and Planning A: Economy and Space* 49, no. 11 (November 2017): 2649–2672. https://doi.org/10.1177/0308518X17723630

Mees, Heleen, and Peter Driessen. "A Framework for Assessing the Accountability of Local Governance Arrangements for Adaptation to Climate Change." *Journal of Environmental Planning and Management* 62, no. 4 (March 21, 2019): 671–691. https://doi.org/10.1080/09640568.2018.1428184

Menikpura, S.N.M., Atsushi Santo, and Yasuhiko Hotta. "Assessing the Climate Co-Benefits from Waste Electrical and Electronic Equipment (WEEE) Recycling in Japan." *Journal of Cleaner Production* 74 (July 2014): 183–190. https://doi.org/10.1016/j.jclepro.2014.03.040

Mookherjee, Dilip, and Pranab Bardhan. "Decentralization, Corruption And Government Accountability: An Overview." *International Handbook on the Economics of Corruption*, July 1, 2005.

Morse, Janice M., and Carl Mitcham. "Exploring Qualitatively-Derived Concepts: Inductive—Deductive Pitfalls." *International Journal of Qualitative Methods*, November 30, 2016. https://doi.org/10.1177/160940690200100404

Mouton, Morgan. "The Philippine Electricity Sector Reform and the Urban Question: How Metro Manila's Utility Is Tackling Urban Poverty." *Energy Policy* 78 (March 2015): 225–234. https://doi.org/10.1016/j.enpol.2014.11.005

Mulgan, Richard. "'Accountability': An Ever-Expanding Concept?" *Public Administration* 78, no. 3 (January 2000): 555–573. https://doi.org/10.1111/1467-9299.00218

Mungkasa, Oswar. "Juggling Jakarta: Urban Challenges and Solutions." Presented at the CLC Lecture Series, CLC, March 13, 2019. https://www.clc.gov.sg/events/lectures/view/juggling-jakarta

Newig, Jens, and Timothy Moss. "Scale in Environmental Governance: Moving from Concepts and Cases to Consolidation." *Journal of Environmental Policy & Planning* 19, no. 5 (September 3, 2017): 473–479. https://doi.org/10.1080/1523908X.2017.1390926

Newman, Peter. "Sustainability and Cities: Extending the Metabolism Model." *Landscape and Urban Planning* 44, no. 4 (1999): 219–226. https://doi.org/10.1016/S0169-2046(99)00009-2

News Desk. "BPTJ Plans to Develop Tram System in Jakarta." *The Jakarta Post*, October 20, 2018, sec. City. https://www.thejakartapost.com/news/2018/10/19/bptj-plans-to-develop-tram-system-in-jakarta.html

Nyawo, Jabulani Christopher. "Public Participation and Accountability in Local Government with Particular Reference to Jozini Municipality." *Journal of Human Ecology* 57, no. 1–2 (February 1, 2017): 60–69. https://doi.org/10.1080/09709274.2017.1311658

Nyoman Wira, Ni. "Blue Bird Launches E-Taxis." *The Jakarta Post*, April 23, 2019, sec. Science & Tech. https://www.thejakartapost.com/life/2019/04/23/blue-bird-launches-e-taxis.html

OECD. "Green Growth in Bangkok, Thailand." *OECD Green Growth Studies. OECD*, August 18, 2015. http://www.oecd.org/governance/green-growth-in-bangkok-thailand-9789264237087-en.htm

Pardede, Giffy. "MEMR Announces Easing of Regulatory Burden: New & Renewable Energy Sector to Benefit." *Lexology*, February 14, 2018. https://www.lexology.com/library/detail.aspx?g=e4f375c2-78ca-4fb9-b2f0-dfcbcb200aa6

Permatasari, Y. I., S. Hardjosoekarto, and R. V. Salomo. "Institutional Transformation of Public Private Partnership to Ensure Low Transaction Costs: Case Study of Infrastructure Provision at the Ministry of Public Works and Housing of the Republic of Indonesia." *Systemic Practice and Action Research* 33 (2020): 149–165.

Phdungsilp, Aumnad. "Energy Analysis for Sustainable Mega-Cities." Doctoral dissertation, KTH, School of Industrial Engineering and Management (Sweden), 2006.

Phdungsilp, Aumnad. "Integrated Energy and Carbon Modeling with a Decision Support System: Policy Scenarios for Low-Carbon City Development in Bangkok." Energy Policy 38, no. 9 (2010): 4808–4817. https://doi.org/10.1016/j.enpol.2009.10.026

Pincetl, Stephanie, Paul Bunje, and Tisha Holmes. "An Expanded Urban Metabolism Method: Toward a Systems Approach for Assessing Urban Energy Processes and Causes." *Landscape and Urban Planning* 107 (September 15, 2012): 193–202. https://doi.org/10.1016/j.landurbplan.2012.06.006

Pollitt, Christopher. *The Essential Public Manager*. 1 edition. Maidenhead: Open University Press, 2003.

Porio, Emma, and Christine Crisol. "Property Rights, Security of Tenure and the Urban Poor in Metro Manila." *Habitat International, Land Tenure and Property Rights* 28, no. 2 (June 1, 2004): 203–219. https://doi.org/10.1016/S0197-3975(03)00068-7

Porio, Emma. "Climate Change Vulnerability and Adaptation in Metro Manila." *Asian Journal of Social Science* 42, no. 1–2 (2014): 75–102. https://doi.org/10.1163/15685314-04201006

Porio, Emma. "Sustainable Development Goals and Quality of Life Targets: Insights from Metro Manila." *Current Sociology* 63, no. 2 (March 2015): 244–260. https://doi.org/10.1177/0011392114556586

Porio, Emma. "Vulnerability, Adaptation, and Resilience to Floods and Climate Change-Related Risks among Marginal, Riverine Communities in Metro Manila." *Asian Journal of Social Science* 39, no. 4 (2011): 425–445. https://doi.org/10.1163/156853111X597260

Prajongsan, Pimolsiri, and Steve Sharples. "Enhancing Natural Ventilation, Thermal Comfort and Energy Savings in High-Rise Residential Buildings in Bangkok through the Use of Ventilation Shafts." *Building and Environment* 50 (April 1, 2012): 104–113. https://doi.org/10.1016/j.buildenv.2011.10.020

Prasetyanto, K., and F. Sari. "Environmental Kuznets Curve: Economic Growth with Environmental Degradation in Indonesia." *International Journal of Energy Economics and Policy* 11, no. 5 (2021): 622–628.

PricewaterhouseCoopers. "Jakarta ERP Project: Progress of Electronic Road Pricing Tender." *PwC*, January 15, 2019. https://www.pwc.com/id/en/media-centre/infrastructure-news/January-2019/jakarta-erp-project-progress-of-electronic-road-pricing-tender.html

Prima Diyarni, Inke. "Low Carbon Development in DKI Jakarta." presented at the Seminar on City to City Collaboration for Low Carbon City Development in Asia, Tokyo, January 30, 2018. https://www.iges.or.jp/files/research/sustainable-city/PDF/20180130/08_DKI_JAKARTA.pdf

Quinlivan, Dale, Margaret Nowak, and Des Klass. "From Accountability to Assurance – Stakeholder Perspectives in Local Government." *Australian Journal of Public Administration* 73, no. 2 (2014): 206–217. https://doi.org/10.1111/1467-8500.12080

Ramaswami, Anu, Abel Chavez, and Marian Chertow. "Carbon Footprinting of Cities and Implications for Analysis of Urban Material and Energy Flows." *Journal of Industrial Ecology* 16, no. 6 (2012): 783–785. https://doi.org/10.1111/j.1530-9290.2012.00569.x

Reinard Sulaiman, Stefanno. "Govt Upbeat over Electric Vehicle, but Charging Stations Still Rare to Find." *The Jakarta Post*, January 21, 2019, sec. Business. https://www.thejakartapost.com/news/2019/01/20/planned-rule-on-electric-vehicles-leaves-questions-on-infrastructure.html

Rohracher, Harald, and Philipp Späth. "The Interplay of Urban Energy Policy and Socio-Technical Transitions: The Eco-Cities of Graz and Freiburg in Retrospect:" *Urban Studies*, August 27, 2013. https://doi.org/10.1177/0042098013500360

Rukmana, D. "Rapid Urbanization and the Need for Sustainable Transportation Policies in Jakarta." *IOP Conference Series: Earth and Environmental Science* 124 (March 2018): 012017. https://doi.org/10.1088/1755-1315/124/1/012017

Sahakian, Marlyne D. "A Matter of Trust in Metro Manila: Collective Action towards 'Green Economy' Transitions." *Development* 55, no. 1 (March 2012): 126–133. https://doi.org/10.1057/dev.2011.114

Sahakian, Marlyne D. "Understanding Household Energy Consumption Patterns: When 'West Is Best' in Metro Manila." *Energy Policy* 39, no. 2 (February 2011): 596–602. https://doi.org/10.1016/j.enpol.2010.10.032

Sahakian, Marlyne D., and Julia K. Steinberger. "Energy Reduction Through a Deeper Understanding of Household Consumption: Staying Cool in Metro Manila." *Journal of Industrial Ecology* 15, no. 1 (February 2011): 31–48. https://doi.org/10.1111/j.1530-9290.2010.00305.x

Santika, W. G., T. Urmee, Y. Simsek, P. A. Bahri, and M. Anisuzzaman. "An Assessment of Energy Policy Impacts on Achieving Sustainable Development Goal 7 in Indonesia." *Energy for Sustainable Development* 59 (2020): 33–48.

Sefriyadi, I., I. G. A. Andani, A. Raditya, P. F. Belgiawan, and N. A. Windasari. "Private Car Ownership in Indonesia: Affecting Factors and Policy Strategies." *Transportation Research Interdisciplinary Perspectives* 19 (2023): 100796.

Serrona, Kevin Roy, and Jeong-Soo Yu. "Finding Urban Waste Management Solutions and Policies: Waste-to-Energy Development and Livelihood Support System in Payatas, Metro Manila, Philippines." *Journal of Environmental Sciences (China)* 21, Suppl 1 (2009): S40–43. https://doi.org/10.1016/S1001-0742(09)60033-4

Shah, Anwar, ed. *Local Governance in Developing Countries*. Public Sector, Governance, and Accountability Series. Washington, D.C: World Bank, 2006.

Shane, A. Megan, and Thomas E. Graedel. "Urban Environmental Sustainability Metrics: A Provisional Set." *Journal of Environmental Planning and Management* 43, no. 5 (September 1, 2000): 643–663. https://doi.org/10.1080/713676586

Shrestha, Ram M., S. Kumar, Samuel Martin, and Arjun Dhakal. "Modern Energy Use by the Urban Poor in Thailand: A Study of Slum Households in Two Cities." *Energy for Sustainable Development* 12, no. 4 (December 1, 2008): 5–13. https://doi.org/10.1016/S0973-0826(09)60003-X

Simanjuntak, J. P., K. A. Al-Attab, E. Daryanto, and B. H. Tambunan. "Bioenergy as an Alternative Energy Source: Progress and Development to Meet the Energy Mix in Indonesia." *Journal of Advanced Research in Fluid Mechanics and Thermal Sciences* 97, no. 1 (2022): 85–104.

Singtong, Saravanee, and Chanathip Pharino. "Using Geospatial Data for Decision Making for a MSW Management Plan in the City: A Case Study in Bangkok, Thailand." *Clean Technologies and Environmental Policy* 20 (2018): 667–680. https://doi.org/10.1007/s10098-018-1495-5

Sipahutar, Tassia. "No Fanfare for Indonesia as Economy Hits Trillion-Dollar Mark." *Bloomberg.Com*, January 3, 2018, sec. Markets. https://www.bloomberg.com/news/articles/2018-01-03/no-fanfare-for-indonesia-as-economy-hits-trillion-dollar-mark

Slaughter, A. M. "The Accountability of Government Networks." In *The Globalization of International Law*. Routledge, 2017, pp. 471–496.

Sovacool, Benjamin, and Marilyn Brown. "Twelve Metropolitan Carbon Footprints: A Preliminary Comparative Global Assessment." *Energy Policy* 38 (September 1, 2010): 4856–4869. https://doi.org/10.1016/j.enpol.2009.10.001

Steinberg, Florian. "Jakarta: Environmental Problems and Sustainability." *Habitat International* 31, no. 3 (September 1, 2007): 354–365. https://doi.org/10.1016/j.habitatint.2007.06.002

Suhari, Syamsul Huda M. "Sulawesi's Largest Solar Power Plant Begins Operation." *The Jakarta Post*, March 31, 2016, sec. National. https://www.thejakartapost.com/news/2016/03/31/sulawesi-s-largest-solar-power-plant-begins-operation.html

Sundaryani, Fedina S. "Regions Told to Draft Energy Plans." *The Jakarta Post*, March 30, 2017, sec. Business. https://www.thejakartapost.com/news/2017/03/30/regions-told-to-draft-energy-plans.html

Surahman, Usep, Osamu Higashi, and Tetsu Kubota. "Evaluation of Current Material Stock and Future Demolition Waste for Urban Residential Buildings in Jakarta and Bandung, Indonesia: Embodied Energy and CO2 Emission Analysis," 2017. https://doi.org/10.1007/s10163-015-0460-1

Syakriah, A. "A Bus Is Open to Everyone Regardless of Class': Riding the World's Biggest Network." *The Guardian*. 13 Nov (2023). <https://www.theguardian.com/environment/2023/nov/30/a-bus-is-open-to-everyone-regardless-of-class-riding-the-worlds-biggest-network>

Terry, James P., Kieran Dunne, and Kruawun Jankaew. "Prehistorical Frequency of High-Energy Marine Inundation Events Driven by Typhoons in the Bay of Bangkok (Thailand), Interpreted from Coastal Carbonate Boulders." *Earth Surface Processes and Landforms* 41, no. 4 (2016): 553–562. https://doi.org/10.1002/esp.3873

The Jakarta Post. "Court Halts Jakarta's Plan to Reopen ERP Tenders." *The Jakarta Post*, March 5, 2020b. https://www.thejakartapost.com/news/2020/03/04/court-halts-jakartas-plan-to-reopen-erp-tenders.html

The Jakarta Post. "New Greater Jakarta Transportation Authority to Be Created in July." *The Jakarta Post*, March 26, 2019. https://www.thejakartapost.com/news/2019/03/26/new-greater-jakarta-transportation-authority-to-be-created-in-july.html

The Jakarta Post. "Odd-Even Policy to Cut Traffic by 2,000 Vehicles." *The Jakarta Post*, March 8, 2018, sec. Business. https://www.thejakartapost.com/news/2018/03/08/odd-even-policy-to-cut-traffic-by-2000-vehicles.html

The Jakarta Post. "Transjakarta to Launch Two Electric Buses in January." *The Jakarta Post*, January 1, 2020a. https://www.thejakartapost.com/news/2020/01/01/transjakarta-to-launch-two-electric-buses-in-january.html

World Bank. *Climate Change Impact and Adaptation Study for Bangkok Metropolitan Region: Final Report.* World Bank, March 1, 2009. http://documents.worldbank.org/curated/en/834261468025503285/Climate-change-impact-and-adaptation-study-for-Bangkok-metropolitan-region-final-report

World Bank. "Data: Indonesia," 2016. https://data.worldbank.org/country/indonesia

Thitanuwat, Bussarakam, Chongchin Polprasert, and Andrew J. Englande. "Green Residues from Bangkok Green Space for Renewable Energy Recovery, Phosphorus Recycling and Greenhouse Gases Emission Reduction." *Waste Management (New York, N.Y.)* 61 (March 2017): 572–581. https://doi.org/10.1016/j.wasman.2016.12.012

Tiangco, M., Alfredo Mahar Lagmay, and J. Argete. "ASTER-Based Study of the Night-Time Urban Heat Island Effect in Metro Manila." *International Journal of Remote Sensing* 29 (May 20, 2008): 2799–2818. https://doi.org/10.1080/01431160701408360

Turner, Jeff. "Urban Mass Transit, Gender Planning Protocols and Social Sustainability – The Case of Jakarta." *World Transit Research*, January 1, 2012. https://www.worldtransitresearch.info/research/4155

UN-HABITAT. "Tools to Support Transparency in Local Governance." Urban Governance Toolkit Series, United Nations Human Settlement Programme (UN-HABITAT), Nairobi, 2004. https://gsdrc.org/document-library/tools-to-support-transparency-in-local-governance/

United Nations. "Sustainable Development Goal 7." *Sustainable Development Knowledge Platform* 2017. https://sustainabledevelopment.un.org/sdg7

Utama, Nuki Agya, and Shabbir Gheewala. "Indonesian Residential High Rise Buildings: A Life Cycle Energy Assessment." *Energy and Buildings - ENERG BLDG* 41 (November 1, 2009): 1263–1268. https://doi.org/10.1016/j.enbuild.2009.07.025

Van Belle, Sara, and Susannah H. Mayhew. "Public Accountability Needs to Be Enforced – A Case Study of the Governance Arrangements and Accountability Practices in a Rural Health District in Ghana." *BMC Health Services Research* 16, no. 1 (October 12, 2016): 568. https://doi.org/10.1186/s12913-016-1836-1

Van Evera, Stephen. *Guide to Methods for Students of Political Science.* 1 edition. Ithaca: Cornell University Press, 1997.

Wafa, Ainul. "Country Report 'Electricity Sector In Indonesia'." *IEEJ*, June 2017, 29.

Wicaksono, A. "Indonesian State-owned Enterprises: The Challenge of Reform." In *Southeast Asian Affairs* (eds.: Daljit Singh and Tin Maung Maung Than), Singapore: ISEAS Publishing (2008): 146–167.

Wongsapai, Wongkot, Ariyah Mekhakul, Manaswee Arayasiri, Kazuhito Yamada, Mariko Fujimori, and Hisaaki Kato. "Developing the City-Level Climate Change Master Plan: The Case of Energy Sector in Bangkok." *MATEC Web of Conferences* 70 (2016): 11001. https://doi.org/10.1051/matecconf/20167011001

Yilmaz, Serdar, and Ayse Guner. "Local Government Discretion and Accountability in Turkey." *Public Administration and Development* 33, no. 2 (2013): 125–142. https://doi.org/10.1002/pad.1646

Yilmaz, Serdar, and Varsha Venugopal. "Local Government Discretion and Accountability in Philippines." *Journal of International Development* 25, no. 2 (2013): 227–250. https://doi.org/10.1002/jid.1687
Yilmaz, Serdar, Yakup Beris, and Rodrigo Serrano-Berthet. "Local Government Discretion and Accountability: A Local Governance Framework." Social Development Working Papers. Washington, D.C.: World Bank, 2007. http://siteresources.worldbank.org/EXTSOCIALDEVELOPMENT/Resources/244362-1193949504055/LocalGovernmentDiscretionandAccountability.pdf
Young, Clara. "Greening Bangkok." *OECD Observer*, 2016. https://oecdobserver.org/news/fullstory.php/aid/5667/Greening_Bangkok.html
Zengerling, Cathrin. "Action on Climate Change Mitigation in German and Chinese Cities – A Search for Emerging Patterns of Accountability." *Habitat International* 75 (May 1, 2018): 147–153. https://doi.org/10.1016/j.habitatint.2018.03.008
Zhang, J., & Li, F. Energy consumption and low carbon development strategies of three global cities in Asian developing countries. *Journal of Renewable and Sustainable Energy* 9, no. 2 (2017).

7

CITY OF CHALLENGES

Toward eco-technological innovation? Examples from Morocco and South Korea

Patrizia Ingallina, Jungyoon Park, and Marco Carlotti

Introduction

The concept of an eco-technological city emerged from research that began around 2008, involving discussions on knowledge-based cities (KBC) and their definition as a field combining urban development, urban studies, and knowledge management (Carrillo, 2006). Unlike smart cities, whose meaning and scope remain to be clarified (Kunzmann, 2014), KBCs are the result of deliberate long-term planning within a broader territorial strategy (Ingallina, 2019a). The idea of a "KBC" and "knowledge-based urban development" (Yigitcanlar *et al.*, 2008; Yigitcanlar, 2011) emphasizes environmental and social aspects alongside technological ones, such as advanced learning techniques and well-equipped educational spaces. A city today cannot be managed without New Information and Communication Technologies (NICT), and how we use them needs to be reviewed and regulated to ensure optimal benefits for everyone – especially in the Anthropocene era (Carrillo, 2021). From this point of view, the concept of an "eco-technological city" summarizes the idea: a city capable of leveraging new technologies as long as they serve humanity.

This has been evident in strategies aiming to create KBCs like the Green City of Benguerir and Technopole of Foum El Oued in Morocco, and renowned initiatives such as the "Knowledge Industrial Green City" (Lee, 2014) at the Magok Industrial Complex in Seoul, Korea (Ingallina, 2017, 2019b). Although Morocco and South Korea represent very different contexts – Morocco being an emerging country recently enhancing R&D through inter-university partnerships (Ingallina, 2019b), and Korea being highly advanced

DOI: 10.4324/9781003439325-10

in technological innovations with the "cluster-oriented strategies" of the Seoul Metropolitan Government since the early 2000s – both are moving toward integrating technological innovations with ecological and social imperatives (Ingallina *et al.*, 2023).

Our primary research in Morocco has advanced the evolving concept of an "eco-technological city," particularly through our direct involvement in the ambitious Technopole of Foum El Oued project in the Greater Laayoune area of the Sahara Desert, strongly supported by the King of Morocco (Benchafai, 2019). This concept was formalized at the *1st International Forum on the City of Challenges: Eco-Technological Innovation* in October 2019, where numerous researchers, professionals, and policymakers discussed the eligibility of the Greater Laayoune area as an eco-technological city. The first project in Benguerir, Morocco, aimed to improve the overall living conditions of a degraded urban and social area, lacking infrastructure and isolated from major traffic flows despite being 70 km from Marrakech. This was achieved through a technologically and ecologically advanced urban project, hoping to trigger positive effects throughout the region.

In Korea, the focus has shifted from a purely "cluster-oriented" approach focused on technological innovation (Sangam Digital Media City (DMC)) to a greener, more inclusive Smart Energy City (Magok Industrial Complex) that considers residents' opinions. Initially labeled as a "knowledge industry green city" (Ingallina, 2019b), Magok eventually adopted the "Smart City" label following the "Act on the Promotion of Smart City Development and Industry (2017)," which defines the characteristics of a smart and sustainable city. This illustrates that while labeling a city is important for defining its characteristics, it is often temporary and influenced by economic goals such as attracting public funding, businesses, tourists, and residents.

The original experiences reported – especially in the case of Morocco – help to understand the efforts made by these two countries to reconcile technological and ecological aspects through "eco-technological cities." The aim of this paper is to shed light on the stated strategies (strategic plans, labels, etc.) materialized in these urban projects and to specify their characteristics. Based on these case studies, questions are asked about their social impacts, to identify gaps and opportunities for future eco-technological urban projects. This work is the result of action-research in which we were involved during visits, the organization of the *1st International Forum on the City of Challenges*, and conducting university training sessions.[1] This chapter is based on the experience from this action-research as well as on a review of existing scientific literature, an analysis of institutional and corporate documents (particularly from promoters), and resources from the local and national press of both countries studied.

The Moroccan case: Integrating eco-technological innovation into new KBCs

In the context of globalization oriented toward innovation, emerging countries are seeing the transition toward knowledge-based economy as a development opportunity. This is the case of Morocco, riding the wave of ICT development since 2000s to foster socio-economic development (Lagzouli et Lakhal, 2020). Furthermore, technological progress has become a strong part of Moroccan strategy to fight climate change and sustainable development challenges (Ministry Delegate of the Minister of Energy, Mines, Water and Environment, in charge of Environment, 2014, p. 21; Kingdom of Morocco, 2017, pp. 45–46).

This orientation toward both technology and ecology is physically manifested through the emergence of "eco-technological cities," trying to redesign a more social and ecological city model. Indeed, Morocco is facing strong socio-economic and demographic challenges. The Kingdom's population has grown considerably over the past 60 years, from 11.6 million in 1960 to 33.8 million in 2014 (Haut-Commissariat au Plan, 2014a).[2] Today, this demographic growth is coupled with an increasing urbanization rate. According to the Haut-Commissariat au Plan (2014b), it increased from 2,901% in 1960 to 6,003% in 2014 across the whole country. This sharp increase has led to a housing crisis in urban areas – caused by limited land availability coupled with very high demand for housing – which is struggling to absorb all these flows, particularly in major urban centers such as Casablanca (Zyani, 2002).

Urban planning in Morocco: From the New Towns Program to a new generation of sustainable cities

NTP as a response to urban issues

From the 2000s onwards, the Moroccan government has endeavored to address urban demographic issues through the New Towns Program (NTP) initiated by the Ministry of Housing and Town Planning (MHU) in 2004. Led by the Al Omrane Group, the program aimed to create 15 new cities by 2020 on the outskirts of Morocco's major cities, addressing the increasing demand for affordable housing for the lower and middle classes (Ministère de l'Aménagement du Territoire National, de l'Urbanisme, de l'Habitat et de la Politique de la Ville, n.d.). Indeed, Morocco experienced a housing critical situation during the 1990s–2000s, especially for the most vulnerable households, which already led King Hassan II to set up the "National Program of 200,000 social housing units per year" in 1994 (Catusse, 2005).

The main objective of the NTP was to urbanize 5,000 hectares to accommodate 1 million people (Boutaleb, 2019), many of whom were living in unsanitary housing and shantytowns. Total investment in the program has exceeded 100 billion Moroccan dirhams, equivalent to 8.8 billion euros (Darras, 2018).

In the end, four towns were officialized by the MHU: Tamansourt, near Marrakesh; Tamesna, in proximity to Rabat; Sahel Lakhyata, at 20 km from Casablanca; and Chrafate, in the Tangier-Tetouan-Al Hoceima region. The remaining cities have not yet been announced until today.

These cities were mediatized and branded around their desirable living environment encompassing large green spaces and quality urban framework (Ballout, 2017). However, these new towns turned into real failures becoming "dormitory cities" with unfulfilled objectives (Harroud & Alillouch, 2022). In 2018, they had 150,000 inhabitants out of the 1 million expected by 2020 (Darras, 2018). Until today, the inhabitants are unsatisfied with their living conditions. Their complaints include inadequate public transport, lack of cultural and sports facilities, and poorly maintained buildings, as reported by Chakir Alaoui and Moussaaid (2022).

Furthermore, Ballout (2017) argues that the green-friendly marketing can be seen more as greenwashing than a genuine commitment to environmental principles. He describes the lack of ecological measures, notably concerning building insulation and transport modes.

The failure of the NTP

The NTP failed by missing strategic planning vision. The Ministry conducted this program with a "top-down vision" of territorial planning, a "traditional" approach to decision-making in which the public body unilaterally decides on the steps to be taken, without any citizen consultation. The absence of citizen involvement in the co-decision process is indeed one of the weak links in Moroccan urban planning (Idrissi, 2014). On this topic, Harroud (2017) criticized the lack of collective discussion and the absence of preliminary feasibility and opportunity studies to conduct the NTP. According to the author, the Ministry has seized land opportunities (cheap land or free land owned by the Kingdom) to rapidly build new towns in order to address the aforementioned problems. However, these location choices were not planned in an optimal and desirable manner. These cities reflect a "catch-up urbanism," driving rapid construction without considering their connection within the national or regional territorial network.

The emergence of a "new generation" of cities: Toward an eco-technological innovation?

Shortly after the dysfunctional NTP, a new generation of towns emerged, attempting to avoid repeating its mistakes (Harroud & Alillouch, 2022). They promote innovative urban planning with a strong emphasis on nature.

The first of these is Zenata Eco-City (Greater Casablanca), launched in 2006 by the Caisse de Dépôt et de Gestion, a public financial institution. It was followed by many other projects from OCP[3]: Green City of Benguerir (2009), Mazagan Urban Pole (2013), Foum el Oued Technopole (2016). OCP is the world leading exporter of phosphate, with the Kingdom of Morocco being the sole investor. As a public company, the company builds new towns to meet national and local socio-economic development goals, especially through knowledge economy. They share a similar model, centered around University Mohammed VI Polytechnic (UM6P) as their core driver of development.

This part highlights the Green City of Benguerir, a new developing town. It represents a "knowledge-based green city," as it integrates high-level educational infrastructure, including a world-class university, as anchor institutions to foster ecological sustainability through technological development. In this way, it could reflect one of the ways of looking at eco-technological innovation through the prism of the knowledge city as a vector for socio-ecological development in a developing country like Morocco.

The Green City of Benguerir, a new green KBC in Morocco

General overview

The Green City of Benguerir is planned to cover an area of 1,300 hectares. It aims to accommodate 100,000 residents, including 25,000 housing units, 200,000 m^2 of retail space, and an 80-hectare green belt by 2045 (OCP, 2022, p. 242).

The city is being built next to the old town of Benguerir, capital of the Rhamna Province. The old city is a medium-sized city with 88,000 inhabitants experiencing high levels of poverty: the city has an unemployment rate of 40% (ASToN Network, 2023). According to Barthel (2015), this site selection was partly influenced by the former president of the Benguerir municipal council. He aimed to establish a city on OCP's land, where an open-pit mine operates. The global underlying objective was to build a town addressing pollution problems caused by OCP-managed extraction sites.

While we don't know the current state of these remediation efforts, OCP has decided to create a "knowledge city of reference, both smart and ecological" (SADV, 2020). To achieve this, the company set up a subsidiary called Société d'Aménagement et de Développement Vert (SADV) in 2011 to design, plan, and develop the overall project (SADV, 2020). Three key points are targeted (OCP, 2022):

- Academic excellence and research through the creation of high-level research institutions;

- Economic development through the creation of an economic activity zone encompassing business centers, start-ups incubator, teleport, data centers, and all the innovation ecosystems necessary – laboratories, academic, and training institutions;
- Urban attractiveness and sustainability with high-quality, smart, and sustainable urban amenities and housing, including a large green space component.

An urban design project based on the knowledge economy

The first phase of the project aimed to create an innovative ecosystem. This is reflected in the first buildings completed, which are knowledge-based infrastructure and researchers' residences (Table 7.1).

The most significant infrastructure is the University Mohammed VI Polytechnic (UM6P) (SADV, 2020). Established in 2013 and inaugurated in 2017, the project was designed by internationally renowned architects Ricardo Bofill and Elie Mouyal. It counts 4,500 students across six campuses nationwide (OCP, 2022, pp. 229–230). UM6P was conceived as an anchor institution for Benguerir and its surroundings. The campus offers an innovative ecosystem offering high value-added trainings (it is specialized on fundamental sciences, applied research, and coding), R&D development, and startup

TABLE 7.1 The three development phases of Benguerir's Green City

	Achieved	*Ongoing*
First phase (2011–2025)	• University Mohammed VI Polytechnic (UM6P) • Green Energy Park • Lycée d'Excellence • Industrial Expertise center • Villas Marguerite • Green & Smart Building Park • 1337 school • Data centers • 1st start-ups incubator	• 2nd start-ups incubator • Student campus • Business center • Health and Care Smart City • Media library
Second and third phases (2025–2045)		

Source: OCP. (2021). *Executive Summary - Sustainability Report 2021*. https://ocpsiteprodsa.blob.core.windows.net/media/2021-10/Executive%20Summary%20-%20Sustainability%20Report%202021.pdf.

support. Through its startup campus StarGate, over 500 African startup projects were supported between 2020 and 2023 (OCP, 2022, p. 234).

The Green City ecosystem also includes the creation of other high-level knowledge infrastructure:

- The Lycée Mohammed VI d'Excellence (LM6E) was built in 2015. Since its creation, it has been the talk of the town with record numbers of its students being admitted to the prestigious France's École Polytechnique: in 2023, 20 students of the 60 places reserved for foreign students passed the competitive entrance exam (Hamri, 2023). LM6E has a social approach by extending its support to the most financially disadvantaged high school students. The school covers the fees for 96% of its students (Kadiri, 2021).
- The Green Energy Park, a research platform dedicated to renewable energies built between the Institute for Research in Solar Energy and New Energies and the UM6P (Green Energy Park, n.d.).
- The 1337 Coding School opened in 2019 offering free and open-to-all computer science training (1337, n.d.).

This knowledge-based ecosystem aims to tackle regional economic realities and Morocco's development challenges, prioritizing sustainability in its growth strategy (OCP, 2022, p. 229). By gathering open-to-all knowledge-based infrastructure (e.g., LM6E, 1337 Coding School, UM6P scholarships), the Green City intends to foster societal development through knowledge, especially in Benguerir facing strong socio-economic disparities. Once trained in high-value-added programs, graduates can seek skilled and better-paid jobs, thus enhancing their standard of living and that of their children. The emergence of new "knowledge workers" fulfills the Kingdom's aspirations for innovation. By increasing the number of graduates in strategic sectors, Morocco aims to enhance its competitiveness and attractiveness, positioning itself as a new regional and global technological hub.

A green project based on smart grids

As a "Green City," the project is placing a strong emphasis on landscaping through the creation of public spaces and green areas. According to OCP (2022), this includes the creation of an 80-hectare green belt along four kilometers, 15 hectares of farming areas, 30 hectares of thalwegs, and the planting of 50,000 trees. Along the green belt, five "oases" are envisioned as focal points for recreational facilities, leisure activities, and social connections, including sports facilities, a media library, a museum, an environmental center, an open-air theater, and more (Challenge, 2014). This landscaping design is strengthened by the implementation of specific guidelines covering architectural, energy, water management, green spaces, and mobility aspects (see Table 7.2).

TABLE 7.2 Features of the Green City of Benguerir

Energy	*Renewable sources*	*The Green City aims to provide all the energy it needs from renewable sources by 2027. Collaboration and partnership are in discussion with OCP Green Energy.*
	Bioclimatic architecture and energy efficiency of buildings	*Urban shapes are designed to optimize the natural circulation of winds and sunshine.*
Green water	Responsible water management system	A responsible water management system that includes double circuit drinking and gray water, rainwater retention, and recycling of wastewater.
	Smart irrigation technology	Partnership with AgriEdge to develop smart irrigation technology to maximize irrigation efficiency.
	Recycled and desalinated water	Partnership in discussion with OCP Green Water to provide recycled water for watering, and desalinated water by 2025.
Transportation	Green belt	A Green belt to structure all the soft routes and connect all districts in a natural way.
	Soft mobilities	• Soft mobility routes (pedestrian and biking paths); • Electrical charges are expected to be implemented around the Green City. Electrical bus project is under study.
Waste	Smart system of waste management	A smart system of waste management, which includes recovery of waste and compost stations, ~~that will be established around 2024~~ to process food scrap and divert waste away from landfills.
Innovation	Semi-autonomous districts	Semi-autonomous districts to guarantee residents close access to shops and services, and helps reduce movement and CO_2 emissions.
	Open innovation promotion	Open innovation promotion: to allow start-ups within the ecosystem to propose innovative solutions to common issues of the city, such as green transportation using electrical bikes.
	Eco-construction	Eco-construction prototype under construction as a research project with UM6P, based on using phosphate waste-rock as raw material, which will allow after the proof-of-concept phase to build housing and facilities.

Source: OCP. (2022). *2022 OCP Sustainability Integrated Report*. Rearranged by Marco Carlotti. https://ocpsiteprodsa.blob.core.windows.net/media/2023-08/OCP_Sustainability_Integrated_Report_2022.pdf.

These requirements enabled the city to acquire LEED-ND (Leadership in Energy and Environmental Design for Neighborhood Development) certification from the U. S. Green Building Council in 2015 (USGBC, 2015). This global certification recognized sustainable development and real estate development projects.

A socio-spatial divide in the making?

While creating a world-class KBC requires building internationally acclaimed infrastructure to enhance its attractiveness, we can still question the social and spatial impacts that this new city might bring. Adjacent to the old town of Benguerir with its traditional compact urban fabric, the Green City stands out due to its extensive land area (1,300 hectares) and large geometrically designed structures (the university campus covers 55 hectares). This spatial disparity is particularly evident in the project's master plan (see Figure 7.1),

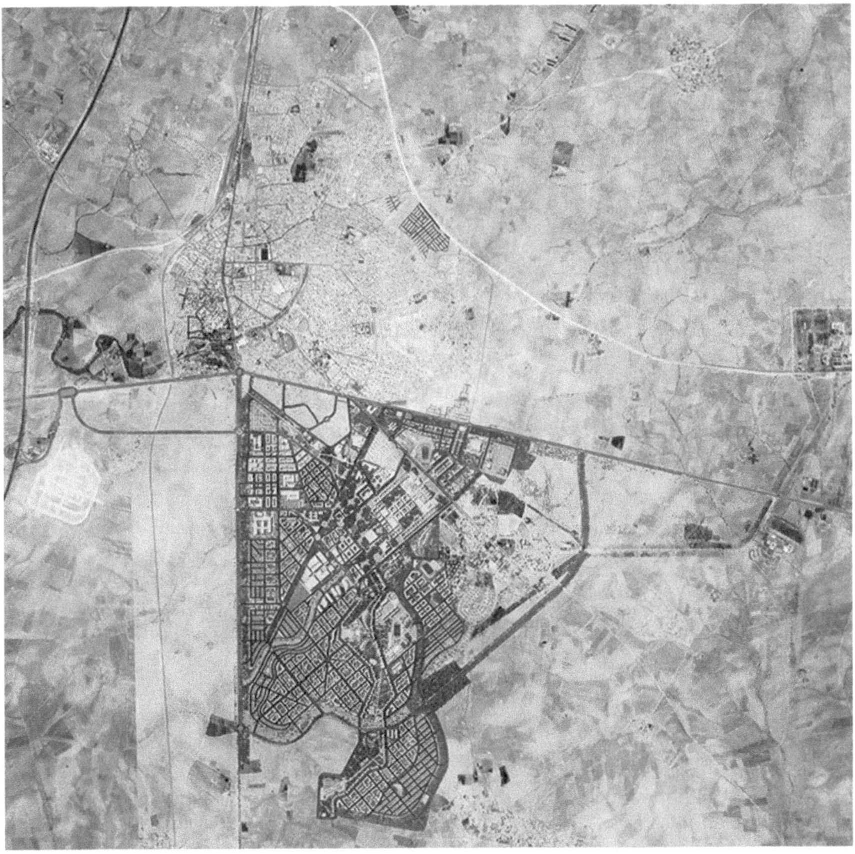

FIGURE 7.1 The Green City of Benguerir.

Source: Sâd Benkirane architecte-urbaniste & MBpaysage, Mounia Bennani paysagiste (n.d.).

which reveals two heterogeneous and distinctive urban fabrics. This spatial configuration prompts the question of interconnection between these two types of cities that are not designed in the same way and inhabited by the same populations.

The impacts of this spatial divide may raise concerns about a potential social boundary between the modern new city and the socio-economically disadvantaged old city. This physical gap could lead to a significant symbolic divide, separating the "wealthy and educated" populations from more vulnerable populations.

Concluding remarks

While the NTP program is struggling and still facing dysfunctions, Morocco is unveiling new ambitions in terms of urban development toward eco-technological cities' development. These new towns are fostering technological innovations to reach socio-economic and ecological sustainability. Mostly driven by OCP, they embody the Moroccan strategy to rise as a regional and global knowledge powerhouse to enhance the overall standard of living of its population. Knowledge is seen as a development lever for the entire population, particularly the most vulnerable, by giving them a chance to improve their standard of living and that of their children. It also enables the achievement of environmental sustainability through the deployment of technological innovations: smart grids, responsible water management system, eco-construction, etc.

However, questions arise in the analysis of these eco-technological projects on spatial and social levels, challenging their ability to fully transition toward a smart and inclusive society in a developing country. Due to their heavy concentration on infrastructure and high-tech solutions, as well as their emphasis on green spaces, these cities may form "ecological enclaves," as asserted by Harroud and Alillouch (2022), but also social and technological ones, potentially leading to concerns about a symbolic boundary between the populations of new and existing cities, resulting in spatial and social fractures.

This is the case with the Green City of Benguerir offering an ultra-modern campus disconnected from the built environment of the old, poor, and degraded city. This spatial configuration carries a risk of social fragmentation between the knowledge workers of the Green City and those of the old city, reflected in terms of living standards and access to technology. Acknowledging these challenges, the city of Benguerir has joined the African Smart Town Network (ASToN) developed by the French Development Agency in 2019 to bridge the socio-economic gap between these two cities. This initiative aims to enhance digital skills among residents through education and the use of digital technologies, thereby enabling them to participate more actively in the local economy (ASToN Network, 2023).

Given these risks, citizen dialogue is crucial to consider the demands of local populations. If a large citizen consultation called "Dessine-moi une ville" (Draw me a city) had been conducted in Foum el Oued Technopole to address the needs of the Sahrawi[4] (Phosboucraa Foundation, 2020), the lack of citizen dialogue in Benguerir could be critical and exacerbate the existing spatial divide.

South Korean Case: Smart city policies and projects in Seoul

Seoul, the capital of South Korea, serves as the nation's administrative and economic hub and has gained international recognition as a leading center of digital technology innovation (Lee, 2023).[5] Since the establishment of the "Seoul Information Comprehensive Plan" in 1999, smart city development has been a key part of the urban development agenda of the Seoul Metropolitan Government (hereafter SMG) for over two decades (SMG Digital Policy Officer, 2024).

This case study aims to primarily examine SMG's Smart City initiatives from the perspective of eco-technological innovation and derive key insights, considering the context of national policies and shifts. For this purpose, two urban design projects, Sangam DMC and Magok Smart City, are analyzed.

National smart city policy framework

Historical review on South Korean smart city policy: from KBC to smart city policy

Since the spatial information database project began in December 2003, led by the Ministry of Construction and Transportation (Lee & Kim, 2007, p. 68), South Korean smart city policies have evolved significantly. The unwavering support from the government has been crucial to these initiatives (Lee *et al.*, 2018). At the beginning of the 21st century, the government adopted the term "Ubiquitous City" (U-City) to reflect its ICT industry development and global connections, marking the attractive first-generation concept of the Korean KBC. The "Act on the Construction, Etc. Of Ubiquitous Cities," passed in 2008, established the legal framework for the U-City policy, providing national support for smart infrastructure primarily for large-scale land development projects exceeding 1.65 million m^2 (Act on the Construction, Etc. Of Ubiquitous Cities, 2009; Lee *et al.*, 2018, p. 39). Thus, its practical application is limited to new city development projects, making it challenging to apply to existing cities (Chung, 2021).

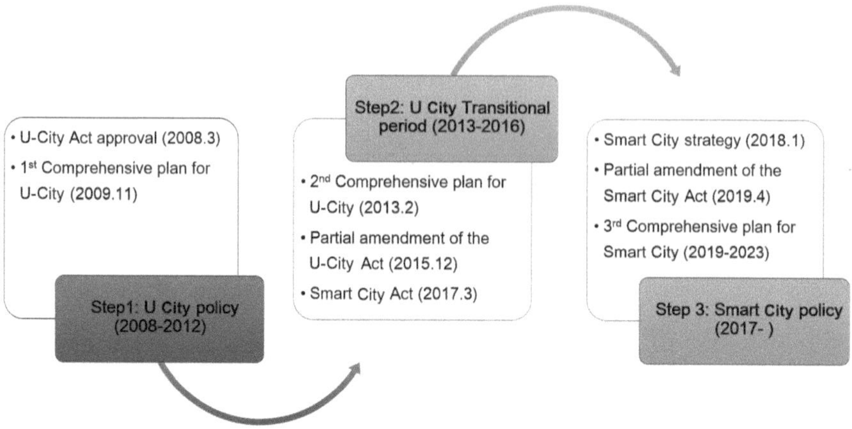

FIGURE 7.2 Historical review on Korean smart city policy framework.

Source: Ministry of Land, Infrastructure and Transport. (n.d.). Smart city Korea official website. Rearranged by Jungyoon Park. https://smartcity.go.kr/.

To overcome this limitation, some improvements were made to the U-City model, shifting focus from new town construction to urban regeneration since the mid-2010s, incorporating environmental and quality of life concepts as well (Chung, 2021). In 2014, the South Korean government established the Second U-City Comprehensive Plan (Lee *et al.*, 2018). During this period, the smart city concept spontaneously spread in the urban policy field in Korea, evolving from the U-City model (LG CNS, 2022).

Following this transitional phase, the term "smart city," reflecting global trends, was officially adopted in 2017 with the implementation of "the Act on the Promotion of Smart City Development and Industry" (hereafter Smart City Act). Article 2 of this act defines a smart city in South Korea as one that pursues sustainability goals through the convergence of technologies to create a more efficient, resilient, and livable urban environment for its residents (Act on the promotion of smart city development and industry, 2017).

Starting in 2018, the South Korean government has been implementing serial initiatives under the Smart City Act (Ministry of Land, Infrastructure and Transport, n.d.). Furthermore, unlike earlier approaches involving only specific ministries, the current smart city policy in South Korea includes participation from all ministries (Lee *et al.*, 2018) (Figure 7.2).

National Support Program for participatory innovation

Experts and public servants may overlook issues like the digital divide, impacting vulnerable groups' access to smart infrastructure (Byun *et al.*,

2018). Neglecting the fundamental principle of smart cities, which addresses urban challenges through ICT for citizens, leads to recurring failures in meeting user needs. This has caused controversy over overlapping investments between public and private sectors, and declining profitability for businesses due to user neglect, jeopardizing sustainability (Cha, 2022).

Since 2018, South Korea has adopted participatory innovation through "Living Labs" as part of its Smart City National Strategy Project, engaging end users in urban regeneration, smart administration, education, and welfare (Moon, 2018). Living Labs involve end users throughout technology development, fostering Public Private People Partnership (PPPP) governance (Seong et al., 2017). For example, the "Digital-based Participatory Program for Local Issues," initiated by the Ministry of the Interior and Safety since 2018, supports over 60 living labs (Ahn, 2022). Our study focuses on the Magok Living Lab in Seoul, showcasing local government efforts in promoting Living Labs.

Smart city projects of SMG

Since the early 2000s, SMG has integrated smart technologies into urban development following the national U-City policy and launched independent smart city projects from 2011 (Lim *et al.*, 2019). Aligned with the Smart City Act, a new master plan was established in 2021, emphasizing the concept of a "smart city" as outlined in the "Smart City & Digitization Master Plan (2021–2025)" (SMG, n.d.).

While Seoul leads in various smart initiatives, concerns arise regarding widening social inequalities due to unequal access to information services, especially among the elderly (Byun *et al.*, 2018). According to the 2022 Digital Information Gap Survey (National Information Society Agency, 2023), the elderly exhibit the lowest level of digital information ability among all digitally vulnerable groups. Seoul, like South Korea overall, grapples with challenges posed by an aging population and low birth rate (*Seoul City Spent 3 Trillion*, 2023). Additionally, as a densely populated city with high economic activity, Seoul faces environmental vulnerabilities such as fine dust, carbon emissions, waste disposal, and high energy consumption (SMG Environment Planning Office, 2023). These concerns are central to Seoul smart city policies, with SMG addressing them through participatory programs like Living Labs (Byun *et al.*, 2018).

Sangam DMC, from U-City exemplary case to a future hub for innovation

Sangam DMC is one of Seoul's prominent urban redevelopment projects from the early 21st century, involving the redevelopment of Nanji Island, which was a landfill site in 1997 (Kim, 2015). Renowned for its role as a media and entertainment cluster, hosting broadcasting companies and ICT firms, it stands as a notable example of Seoul's intelligent city development.

Overview of Sangam DMC project

Sangam DMC is an "Art Digital Media Entertainment (M&E) Cluster" located in Sangam-dong, Mapo District, Seoul, with an area spanning 569,925 m^2 (SBA, 2009, p. 15). Conceived in the late 1990s, construction began in 2002 and concluded in 2015 (Seoul Institute, 2017). It is part of the "Sangam New Millennium City" initiative, chosen to host the 2002 World Cup, leading to the construction of the main stadium (Byun *et al.*, 2018) (Figure 7.3).

Originally one of the most deprived areas of the city, the Sangam New Millennium City site was a former waste disposal site known as "Nanjido," now transformed into the "World Cup Park," a complex of ecological parks (Seoul Institute, 2017). The DMC project, led by the SMG with various organizations' assistance, ensured timely completion of essential infrastructure like the World Cup stadium and subway station (Kim, 2015). Following negotiations between the national government and Seoul, major companies like Samsung Electronics, Hyundai Mobis, and LG CNS joined the project, boosting growth and investment (Kang, 2010). By January 2016, Sangam DMC housed 870 companies with a 94% occupancy rate, employing 40,090 people, an increase of 87 companies and 6,215 employees since 2013 (SBA, 2016).

FIGURE 7.3 Sangam DMC.

Source: Jungyoon Park.

Sangam DMC was initially designed as a knowledge-based cluster to attract advanced technology and innovative industries, focusing on corporate attraction and technological development. As a result, its promotion strategies lack emphasis on sustainability, ecology, and inclusivity, which are key elements of the current smart city agenda.[6] However, though Sangam DMC aligns with the technology centric goals of typical U-City policy, it is not solely confined to past agendas. It is also evolving with the Fourth Industrial Revolution, necessitating investments in high-level technologies like AI, autonomous driving, digital twins, and IoT. Since the enactment of the Smart City Act in 2018, SMG, in collaboration with the Ministry of Science and ICT, signed a memorandum of understanding for the "Advanced ICT-based Sangam DMC Activation Cooperative Project," jointly pursuing 5G and IoT-related initiatives (Byun, *et al.*, 2018). Furthermore, as part of the "Next-Generation Intelligent Transport System (C-ITS) Demonstration Project" by the Ministry of Land, Infrastructure and Transport, Sangam DMC was selected as a 5G-integrated test bed for autonomous driving and also serves as a test bed for Seoul's smart road infrastructure (Chung, 2019).

Magok, new smart energy city in Seoul

Magok district is situated in the western Han River basin of Seoul and is renowned as Seoul's last industrial district. It is a downtown-type new city that has seen urban renewal projects since the early 2000s. Substantial regeneration efforts commenced in 2007, with a total area of 3,655,722 m^2, including 595,340m^2 designated for residential purposes, alongside advanced industrial complexes, public facilities, and the Seoul Botanical Garden (Byun *et al.*, 2018, p. 38).

Overview of Magok development project

The recent Magok district redevelopment project was carried out through a collaboration between the SMG and SH Corporation. SMG was responsible for defining the spatial scope, establishing development and management plans, and approving implementation plans, while overseeing the corporate attraction strategy. SH Corporation (a subsidiary of the SMG), as the implementing agency, was responsible for tasks such as housing construction, sales and land disposition, and the development and implementation of plans. Here is some information about Magok (SMG Balanced Development Office, 2024):

- The Magok development program (about 3,669,000 m^2) was divided into three sub-zones: residential development (1,066,000 m^2), industrial and

business district development (1,904,000 m²), and the installation of a botanical garden and lakeside park (697,000 m²);
- Location: Magok-dong, Gangseo-gu;
- Total Cost: KRW 7,249.6 billion;
- Duration: 2007–2022 ((as of September 2024, the completion rate for Phases 1 and 2 is 99.9%) (Figure 7.4).

The U-City act allowed the inclusion of smart infrastructure from the conception stage of the project; however, in 2018, the Smart City Act introduced a new "Smart Energy City Development Project" (Lim *et al.*, 2019). The project outlines a smart energy model with elements such as microgrids, solar power, and an integrated energy management system in five key areas (Lim *et al.*, 2019):

- Smart energy home: IoT-enabled remote energy management for home devices
- Smart energy building: Eco-friendly energy efficiency management system for buildings
- Smart energy community: Energy-sharing economy in apartment complexes
- Smart energy town: Solar power profit sharing and peer-to-peer energy trading in the community
- Smart heat network: Heating systems using unused and renewable energy
- Policies to solve social problems: the digital divide, inequality issues, and municipal government efforts

The Magok Project is actively advancing a range of eco-technological innovations, including programs centered on energy transition, intelligent waste

FIGURE 7.4 Magok District.

Source: Jungyoon Park.

management, and low carbon smart mobility. The expected benefits of the Magok Smart Energy City project include (Youn, 2018):

- Attaining a +30% self-sufficiency rate in electricity;
- Realizing an annual reduction of 190 tons in fine dust;
- Achieving an annual decrease of 180,000 tons in CO_2 emissions.

The project, initially aimed for completion by 2022, has not received official confirmation from the SMG regarding goal attainment by the end of 2023.[7] Additionally, the Magok Combined Heat and Power Plant project, led by the SMG, faces persistent delays due to opposition from local residents. Differing perspectives exist, with some seeing the plant as a significant environmental concern, while others argue its necessity for achieving carbon neutrality and addressing local heat demand (Park, 2018).

Lastly, Magok district is noteworthy for its Smart City Living Lab projects engaging local residents. In 2019 and 2020, SMG conducted ten Living Labs through open competitions, fostering public-citizen-private sector collaborations. Initiatives included a "Fire Situation Awareness System," "Outdoor Delivery Testing with Autonomous Driving-Based Robots," "Inclusive City for the Visually Impaired," "Community Mapping for Odor in Magok Smart City," etc. (Ministry of Land, Infrastructure and Transport, 2019). Magok Living Lab stands out for its inclusive approach, utilizing appropriate technology to improve citizens' daily lives without solely focusing on extensive infrastructure projects.

Concluding remarks

The strategy to pursue an eco-technological city, defined as an urban design approach that harmonizes technological innovation and ecological sustainability (Ingallina, 2019b), is evident in the transition from the U-City model to the smart city model in South Korea. This shift in urban policy paradigm can be observed by contrasting the Sangam DMC case, which pursues a competitive ICT cluster, with the Magok case, which aims to be a citizen-participatory smart energy city.

However, persistent challenges hinder the development of integrated city platforms. Factors such as the concentration of businesses in the capital region, a predominantly public-driven economy, a lack of voluntary collaboration among private companies, and regulatory constraints on sharing private data limit progress (Kim *et al.*, 2019, p. 116). Furthermore, transitioning into a smart society requires more than just large-scale urban projects like Sangam DMC or Magok district. Transforming existing cities into smart cities is crucial (Jeon, 2022), especially in South Korea, where limited space and stagnant population growth make new town projects challenging (Choi, 2019).

Additionally, developing new towns often raises sustainability concerns by consuming natural greenery and agricultural land.

Last but not least, it is crucial to address the digital divide issue. While public smart city platforms and remote services advance smart city goals, they risk excluding digitally vulnerable groups (Cho & Lee, 2021). Strategically deploying digital technologies can promote inclusive policies, benefiting all citizens, as leading global cities actively pursue inclusive smart city development for marginalized groups (Lee, 2022). Such efforts are also evident in the Sangam and Magok districts analyzed in this paper. Among the Magok living lab experiments in 2019, there is also an app service designed to be user-friendly for transportation-disadvantaged individuals (Lah, 2019). Since September 2022, the Seoul Smart City Center in Sangam DMC has supported digital novices (Cho, 2022). Digital inclusion policies are pursued at various levels, from local districts like Magok and Sangam to regional and national governments. In 2020, the government established the "Digital Inclusion Promotion Plan," encouraging SMG to launch a comprehensive plan for digital capabilities through education and accessibility enhancements (SMG, 2020).

Now, South Korea must create an environmentally friendly urban environment that benefits all social groups, encompassing both newly developed cities and existing urban areas. Addressing these challenges is crucial for creating an inclusive eco-technological city.

General conclusions

Through this study, the concept of "eco-technological innovation," which integrates KBCs, smart technology, and ecological goals, is examined within the context of territorial policy and urban design projects in two distinct countries. In both cases, a strong political will to create economically competitive clusters aimed at attracting knowledge workers (universities and high-value-added companies) is evident in planning strategies that utilize broad communication tools. The success of these projects relies on stakeholder involvement, their support, and effective communication with residents.

In Morocco, the attractiveness of new territories is branded with ecological labels like "Green City" and "Eco-City." In South Korea, the focus is on the "smart" aspect, emphasizing industry connectivity and public infrastructure (DMC, Magok Smart City). These concepts evolve with government decisions and socio-economic changes. The projects are not entirely linear and exhibit contradictions. For example, the Green City of Benguerir presents a stark contrast with the nearby small village, both spatially and socially. South Korea's DMC and Magok projects face challenges such as an aging population, overcrowded urban areas, and high real estate costs.

Furthermore, while economic outcomes like attracting businesses and residents are easily verified, assessing environmental and social achievements is challenging due to the lack of publicly available data. Much of the information is confidential or found in official sources, highlighting a limitation of the study.

Notes

1 A training program from the University of Paris-Sorbonne titled "Ecological Innovation in Architectural and Urban Projects for the Development of the Southern Regions of Morocco" was conducted in 2017 for the engineers and architects responsible for constructing the Technopole of the City of Knowledge in Laayoune.
2 2014 is the year of the latest Moroccan census carried out by the Haut-Commissariat au Plan (HCP), Morocco's official statistics producer. The HCP forecasts a population of over 43.5 million by 2050.
3 Formerly Office Chérifien des Phosphates.
4 This citizen consultation was organized by the Phosboucraa Foundation of OCP Group. Its main objective was to assess the specific needs of children, young people, woman and people with reduced mobility.
5 Seoul was recognized as the world's best smart city by winning the Best City Award in the urban strategy category at the Smart City Expo World Congress (SCEWC) 2022 (Lee, 2023).
6 As discussed above, this project aimed to redevelop the former Nanjido landfill site, by integrating the eco-friendly World Cup Park. However, most city government reports indicate that Sangam DMC's main focus is on establishing a "media & entertainment" and "ICT" industrial cluster, along with developing high-quality residential complexes.
7 Kim Seyong, former CEO of Seoul Housing & Communities Corporation during the early stages of the Magok Smart Energy project, revealed that the project is still ongoing to achieve these goals at the 38th Future Land Forum held by the National Academy of Engineering of Korea (NAEK) on December 14, 2023.

References

1337. (n.d.). *The 1337 Dna*. https://www.1337.ma/en/

Act on the Construction of Ubiquitous Cities, 2009 (S. Kor.), translated in KLRI (Korea Legislation Research Institute)'s online database, https://elaw.klri.re.kr/kor_service/lawView.do?lang=ENG&hseq=17134

Act on the Promotion of Smart City Development and Industry, 2017 (S. Kor.), translated in KLRI's online database. https://elaw.klri.re.kr/kor_service/lawView.do?lang=ENG&hseq=46605

Ahn, H. (2022, May 12). The Ministry of the Interior and Safety Promotes a Digital Innovation Project for Residents [in Korean]. *Etnews*. https://www.etnews.com/20220512000237

ASToN Network. (2023, January 24). Ben Guerir développe l'e-santé et favorise l'employabilité des jeunes. *Medium*. https://aston-network.medium.com/ben-guerir-d%C3%A9veloppe-le-sant%C3%A9-et-favorise-l-employabilit%C3%A9-des-jeunes-c75b1b8fb579

Ballout, J.-M. (2017). Un bilan intermédiaire du *Programme de villes nouvelles* au Maroc. *Les Cahiers d'EMAM, 29*. https://doi.org/10.4000/emam.1316
Barthel, P.-A. (2015). Le Maroc à l'heure de l'éco-urbanisme : innovations et décalages sur le terrain des premiers projets urbains « verts ». In B. Dupret, Z. Rhani, A. Boutaleb, & J.-N. Ferrié (Eds.), *Le Maroc au présent* (pp. 77–89). Centre Jacques-Berque. https://doi.org/10.4000/books.cjb.1001
Benchafai, A. (2019, October 18). *La contribution du SDAU du Grand Laayoune à la fabrication d'une métropole compétitive et durable dans le Sahara Marocain* [Conference presentation]. 1st International Forum on the City of Challenges: Eco-Technological Innovation, Laayoune, Morocco.
Boutaleb, R. (2019, December 12). Programme Villes Nouvelles : Les leçons d'un échec programmé. *Challenge*. https://archive.challenge.ma/programme-villes-nouvelles-les-lecons-dun-echec-programme-125483/
Byun, M., Kim, M., Yi, Ch., & Park, M. (2018). *Social Issues in the Smart city era and the SMG*. Seoul Institute.
Carillo, F. J. (2021). Introduction to Knowledge For The Anthropocene. In F. J. Carrillo, & G. Koch (Eds.), *Knowledge for the Anthropocene. A Multidisciplinary Approach* (pp. 1–9). Edward Elgar Publishing. https://doi.org/10.4337/9781800884298.00009
Carrillo, F. J. (Ed.). (2006). *Knowledge Cities. Approaches, Experiences, and Perspectives*. Routledge.
Catusse, M. (2005). Les réinventions du social dans le Maroc "ajusté". *Revue des mondes musulmans et de la Méditerranée*, 105–106, 175–198.
Cha, J. (2022, September 17). *Expansion of Smart city Living Lab* [in Korean]. The Information and Communication Newspaper. https://www.koit.co.kr/news/articleView.html?idxno=102940
Chakir Alaoui, M., & Moussaaid, B. (2022, September 24). Modèle ou fiasco? Visite guidée de Tamesna, ville nouvelle créée il y a 15 ans à a périphérie de Rabat. *Le 360*. https://fr.le360.ma/societe/modele-ou-fiasco-visite-guidee-de-tamesna-ville-nouvelle-creee-il-y-a-15-ans-a-la-peripherie-de-267469/
Challenge. (2014, July 17). *La nouvelle ville de Benguérir, la première ville verte en Afrique*. https://archive.challenge.ma/la-nouvelle-ville-de-benguerir-la-premiere-ville-verte-en-afrique-30903/
Cho, K., & Lee, Y. (2021). *Research on Elderly Support Measures for Smart Welfare in Seoul* [in Korean]. Seoul 50+ Foundation.
Cho, S. (2022, September 22). Kiosk No Longer Difficult at Seoul Smart City Center! [in Korean]. *Seoul Media Hub*. https://mediahub.seoul.go.kr/archives/2005635
Choi, B. (2019, May 12). The Constraints of New Town Development in the Capital Region [in Korean]. *Hankyoreh*. https://www.hani.co.kr/arti/opinion/column/893578.html
Chung, H. (2021). The Status and Implications of Smart Cities in Korea [in Korean]. *KDB Monthly*. 792: 31–47.
Chung, K. (2019, May 28). 5G Smart City, Sangam DMC as a 'Vital Hub' [in Korean]. *Newspim*. https://www.newspim.com/news/view/20190528000755
Darras, R. (2018, March 22). Villes Nouvelles au Maroc : dix ans après, des objectifs non atteints. *Jeune Afrique*. https://www.jeuneafrique.com/mag/540954/economie-entreprises/villes-nouvelles-au-maroc-dix-ans-apres-des-objectifs-non-atteints/
Green Energy Park. (n.d.). *Green Energy Park*. https://www.greenenergypark.ma/

Hamri, S. (2023, August 5). Prépas scientifiques: Le lydex, potentiel rival des grands lycées parisiens ? *Medias24*. https://medias24.com/2023/08/05/prepas-scientifiques-le-lydex-potentiel-rival-des-grands-lycees-parisiens/

Harroud, T. (2017). Handicaps et contradictions du Programme de villes nouvelles au Maroc. *Les Cahiers d'EMAM*, 29. https://doi.org/10.4000/emam.1333%20Top%20of%20page

Harroud, T., & Alillouch, R. (2022). The "new generation" of New Cities in Morocco: from the "dormitory" City to the "sustainable" City, What has Changed? *African and Mediterranean Journal of Architecture and Urbanism*, 4(2). https://doi.org/10.48399/IMIST.PRSM/amjau-v4i2.36201

Haut-Commissariat au Plan. (2014a). *Recensement Général de la Population et de l'Habitat*. http://rgphentableaux.hcp.ma/Home/

Haut-Commissariat au Plan. (2014b). *Taux d'urbanisation*. https://www.hcp.ma/Taux-d-urbanisation_a3243.html

Idrissi, A. (2014, June 16–21). *Urban Planning in Morocco Between the Theoretical Basics and Legal Principles* [Conference presentation]. FIG Congress 2014, Kuala Lumpur, Malaysia.

Ingallina, P. (2017). Knowledge Cities: Concepts and practices. The Korea and Morocco cases: towards an ecological - industrial knowledge city model?. In G. A. Larrea (Ed.), *Proceedings of the Tenth Congress of Knowledge Cities Worlds – KCWS 2017* (pp. 13–42). Knowledge Cities World Summit.

Ingallina, P. (Ed.). (2019a). *Ecocity, Knowledge city, Smart city. Vers une ville éco-soutenable ?*. Presses Universitaires du Septentrion.

Ingallina, P. (Ed.). (2019b). Vers une Knowledge Ecological Innovation City: les cas de la Corée et du Maroc. In *Ecocity, Knowledge City, Smart City. Vers une ville écosoutenable?* (pp. 205–224). Presses Universitaires du Septentrion.

Ingallina, P., Park, J., & Carlotti, M. (2023, December 12). *City of Challenges: Towards Eco-Technological Innovation? Examples from Morocco and South Korea* [Conference presentation]. 2023 United Nations Climate Change Conference, Dubai, United Arab Emirates.

Jeon, Ch. (2022, Sep 13). Transforming Existing Cities into Smart Cities for Citizen-led Smart Communities [in Korean]. *Engineering Journal*. http://www.engjournal.co.kr/news/articleView.html?idxno=2003

Kadiri, G. (2021, April 27). Le Lycée d'excellence de Benguérir, fabrique marocaine des futurs polytechniciens. *Le Monde*. https://www.lemonde.fr/afrique/article/2021/04/27/le-lycee-d-excellence-de-benguerir-fabrique-marocaine-des-futurs-polytechniciens_6078280_3212.html

Kang, J. (2010, May 10). Sangam DMC, Media and Communication Driven IT Mecca [in Korean]. *Digital Times*. https://www.dt.co.kr/contents.htm?article_no=2010051102011260739001

Kim, D. (2015). Seoul's Digital Media City. *Seoul Solution*. https://www.seoulsolution.kr/en/content/seoul's-digital-media-city-dmc

Kim, I., Lee, J., Seo, Y., Lee, J., Jung, M., & Kim, B. (2019). *A Study on Innovation Ecosystem Promotion in the Korean Smart Cities* [in Korean]. Korea Research Institute for Human Settlements (KRIHS).

Kingdom of Morocco. (2017, October). *Stratégie Nationale de Développement Durable 2030*.

Kunzmann, K. R. (2014). Smart Cities: A New Paradigm of Urban Development. *CRIOS*, 7.

Lagzouli, A., & Lakhal, T. (2020). Knowledge Economy in the Moroccan Context: Theoretical Aspects and Measurement Indicators. *International Journal of Accounting, Finance, Auditing, Management and Economics*, *1*(1), 99–117.

Lah, H. (2019, March 2). Living Lab in Magok Smart City: Making Transportation Accessibility Easier with Smartphones [in Korean]. *Eroun net*. https://www.eroun.net/news/articleView.html?idxno=7465

Lee, B. & Kim, K. (2007). *The Local Government Strategies fort the Realization of u-City* [in Korean]. Korea Research Institute for Local Administration.

Lee, J. (2022). *Small Yet Resilient Smart Cities Contributing to Balanced Local Development* (KR). Smart City Korea. https://smartcity.go.kr

Lee, J., Lee, M., Lee, J., Kim, I., Lee, S. & Young, J. (2018). *A Study on Strategies by Type of Smart City*. KRIHS.

Lee, K. (2014, July 7). *Magok: 'Green Next-generation Industrial R&D Complex in the Heart of City'*. Korea Post. https://www.koreapost.com/news/articleView.html?idxno=592

Lee, N. (2023, March 2). Seoul, Standing as a Leading Digital City... to Learn Expertise, Heads Overseas [in Korean]. *ZDNET Korea*. https://zdnet.co.kr/view/?no=20230301193032

LG CNS. (2022, November 16). The Data Moves the City? Evolution from 'U-city' to 'Smart city'. *LG CNS Website*. https://www.lgcns.com/blog/cns-tech/smartcity/37208/

Lim, H., Ju, J., Yoon, S., Yu, J., Han, Y. & Jin, H. (2019). *A Study on the Diagnosis of Major Projects and Future Plan of Smart City in Seoul*. Seoul Institute.

Ministère de l'Aménagement du Territoire National, de l'Urbanisme, de l'Habitat et de la Politique de la Ville. (n.d.). *Villes et Zones d'Urbanisation Nouvelles*. https://www.mhpv.gov.ma/fr/villes-et-zones-durbanisation-nouvelles/

Ministry Delegate of the Minister of Energy, Mines, Water and Environment, in charge of Environment. (2014, March). *Moroccan Climate Change Policy*.

Ministry of Land, Infrastructure and Transport. (2019). *SMG Announced the Outcome of Magok Smart City Living Labs*. Smart City Korea. https://smartcity.go.kr

Ministry of Land, Infrastructure and Transport. (n.d.). *Laws & Systems of Smart City*. Smart City Korea. https://smartcity.go.kr

Moon, B. (2018, March 30). Two Local Governments Selected to Pilot the Smart City National Strategy Project [in Korean]. *Etnews*. https://www.etnews.com/20180329000163

National Information Society Agency. (2023). *2022 Digital Information Gap Survey*. https://www.nia.or.kr/site/nia_kor/ex/bbs/View.do?cbIdx=81623&bcIdx=25353&parentSeq=25353

OCP. (2021). *OCP Sustainability: Executive Summary*. https://ocpsiteprodsa.blob.core.windows.net/media/2021–10/Executive%20Summary%20-%20Sustainability%20Report%202021.pdf

OCP. (2022). *OCP Sustainability Integrated Report*. https://ocpsiteprodsa.blob.core.windows.net/media/2023-08/OCP_Sustainability_Integrated_Report_2022_0.pdf

Park, K. (2018, May 17). Easing Distrust, Improving Profitability, Long road for Combined Heat and Power Generation [in Korean]. *Industry News*. https://www.industrynews.co.kr/news/articleView.html?idxno=23800

Phosboucraa Foundation. (2020). *« Dessine-moi une ville », une concertation citoyenne, Février 2020*. https://www.technopolefoumeloued.ma/fr/actualite/dessine-moi-une-ville-une-concertation-citoyenne-fevrier-2020

Sâd Benkirane architecte-urbaniste, & Mbpaysage, Mounia Bennani paysagiste. (n.d.). [Green City of Benguerir Urban Development Plan]. https://www.mbpaysage.ma/Pages/detailsProjet?k=38

SADV. (2020). *Green City of Benguerir. A Forward-Looking Project Open to the World.* https://sadv.ma/sites/default/files/2020-10/OCP_Brochure%20SADV_VilleVerte_VAng_Couv3Volets_22x22cm%20V2.pdf

SBA. (2016). Sangam DMC: 94% Office Occupancy Rate, Over 40,000 Employees ! [in Korean]. *SBA Official Blog.* https://blog.naver.com/seoulsba/220724434495

SBA (Seoul Business Agency). (2009). *Digital Media City. Dreams Made a Reality! SMG (Seoul Metropolitan Government)* [in Korean].

Seong, J., Song, W., Jung, B., Choi, C., Youn, C., Jung, S., & Han, K. (2017). Current Status of Korean Living Labs and Its Development Plan. *Policy Res*, 9, 1–194.

Seoul City Spent 3 Trillion Won but Birthrate Still Declined... [in Korean]. (2023, April 27). *Newspim.* https://www.newspim.com/news/view/20230427000573

Seoul Institute. (2017). *Promoting DMC Phase 2 as a leading industry hub* [in Korean]. SMG.

SMG. (2020, October 12). *SMG Digital Capacity Enhancement Plan in the Era of COVID-19* [Press Release] [in Korean]. https://www.seoul.go.kr/news/news_report.do#view/325740

SMG. (n.d.). *Smart City & Digitization Master Plan.* SMG official website. https://english.seoul.go.kr/policy/smart-city/vision-governance

SMG Balanced Development Office. (2024, December 24). *Magok project progress by district* [in Korean]. SMG News Site. https://news.seoul.go.kr/citybuild/archives/65286

SMG Digital Policy Office. (2024, January 25). *Comprehensive plans for Digital Seoul* [in Korean]. SMG News Site. https://news.seoul.go.kr/gov/archives/2092

SMG Environment Planning Office. (2023, October 20). *2050 Seoul Climate Action Plan* [in Korean]. SMG News Site. https://news.seoul.go.kr/env/archives/526703

USGBC. (2015, February). *Ville Verte Mohammed VI.* https://www.usgbc.org/projects/ville-verte-mohammed-vi

Yigitcanlar, T. (2011). Redefining knowledge-based urban development. *International Journal of Knowledge-Based Development*, 2(4), 340–356.

Yigitcanlar, T., Velibeyoglu, K., & Baum, S (Eds.). (2008). *Knowledge-Based Urban Development: Planning and Applications in the Information Era.* IGI Global.

Youn, J. (2018, July 23). Creating a Smart Energy City in Magok District [in Korean]. *Aju News.* https://www.ajunews.com/view/20180723133717938

Zyani, B. (2002). Habitat, contrainte foncière et développement urbain à Casablanca. In P. R. Baduel (Ed.), *Habitat, État, société au Maghreb.* CNRS Éditions. https://doi.org/10.4000/books.editionscnrs.811

8

SMARTIFYING [INFORMAL] CITIES THROUGH PLATFORM URBANISM

Lynn Abdouni

Introduction

Definitions of smart cities have proliferated in the last two decades, developing a multifaceted meaning of what constitutes 'smart city'. Descriptions have evolved to include qualities of people and communities as well as Information and Communication Technology (Albino, Berardi, and Dangelico 2015, 3). A smart city can be described as the result of human knowledge and creative strategies aiming to support the socio-economic and logistic performance within the city through technological infrastructure (Kourtit & Nijkamp, 2012, 93). Platform urbanism, a more recent concept, addresses the relationship between city transformation and digital platforms that mediate everyday services such as transportation (mediated by ride-hailing and delivery applications), navigation (mediated by Google maps, Waze, or similar applications), shopping (mediated by online grocery shopping or e-commerce applications), sometimes involving input from other city dwellers, as discussed later in this chapter.

The informal built environment encompasses unregulated, self-organized urban spaces, often created out of necessity in response to economic or systemic constraints.

This chapter summarizes an exploratory study into how platform urbanism could offer paths for informal built environments to become 'smart'. The chapter begins by defining these concepts in greater depth, followed by an analysis of delivery application interfaces. It concludes with a discussion on how platforms can begin drafting a blueprint that bridges service gaps in informal built environments, potentially paving an alternative pathway to smart cities.

DOI: 10.4324/9781003439325-11

Platform urbanism and the smart city

Conceptually, *platform urbanism* and *smart cities* overlap and may even be considered theoretical extensions of one another, although there are distinct differences between the two. The core of platform urbanism is to provide a mediating lens between city dwellers (or platform users) and the urban systems that surround them (Caprotti et al., 2022). The essence of a smart city is to enhance or improve the quality of life in the city through digital technologies, communication technologies, and data analytics, with applications in transportation, environment, energy, safety and security, education, healthcare, and governance (Abdi & Shahbazitabar, 2020; Gracias et al., 2023).

While the rationale of a smart city is to enhance its existing physical infrastructure with technology, that of platform urbanism is to provide a digital portal through which the citizen experiences the city. This is further articulated in differences pertaining to the temporality, governance, and spatiality aspects. Smart cities imply a long-term geographically constrained effort driven by city authorities, whereas platform urbanism is a short-term, geographically flexible effort operated by technology firms (Caprotti et al., 2022).

Platform urbanism emerged from a conceptual permutation between platform capitalism and smart city planning. Platform capitalism establishes the universe of platforms, exemplified in smartphone and online applications, as a distinct form of capitalism. Although it is subject to a changing uncertain set of rules, and has in recent history generated disruptors such as Uber, Amazon, and similar platforms, it evidently upholds the main functions of a capitalist system (Srnicek, 2017). The smart city strives to enhance the dweller's life through a hybridized system of physical and digital infrastructure. The propagation of a platform market that digitally mediates the daily urban life, including transportation, goods and service purchase, and access to health and education, has been instrumental in bringing cities into the 'smart city' classification. That same market bears all the indications of platform capitalism.

Efforts in smart city planning are often ambitious and large-scale. A few notable examples include the Smart Cities Mission in India, a multi-scalar governmental effort promising to revamp existing transportation and utility infrastructure and leveraging an extensive digital network for livable cities. This is also accompanied by capacity-building, centering the human experience to any development efforts (smart cities, India). Saudi Arabia's NEOM proposes a futuristic city centered around business development in 14 sectors, livability, and nature conservation. Launched in 2017, the project began developing in distinct regional phases with particular attention to branding and sustainable infrastructure (Neom). Across the board, smart cities seek to harness the potential of data, communication technologies, and analytics to enhance various aspects of urban life, including transportation, environmental sustainability, energy efficiency, safety, education, healthcare, and governance. While the smart city movement decidedly indicates a global desire for advancement and livability, it is crucial to note that these efforts are often centered around cities with existing infrastructure and governmental support. This leaves the question of how such smart city approaches could be adapted to or implemented in informal, unplanned, or minimally planned urban environments.

City planning efforts, including but not limited to smart city planning, often circumvent informal urban environments, citing the lack of clear governance, viable infrastructure, and formal regulations as impediments to development. Indeed, the conventional assumption is that planning effort requires a capable governing entity to conceive, orchestrate, and implement plans. This exclusion perpetuates cycles of inequity, as the benefits of smart planning remain elusive for populations living in informal settlements. Addressing the limitations entails reevaluating the significance of informality, starting with centering historical and local knowledge in city making and developing bottom-up approaches that empower dwellers to bring the benefits of smart planning to these underserved areas (Roy, 2016).

The first step to leverage local knowledge and city-making is constructing the framework to identify shortcomings and allocate priorities, and is also identifying site-specific pathways toward implementation. This often involves a lengthy and challenging process rooted in participatory planning that enhances adaptability and stability and thus enables shaping the process into a continuous learning system (Smith, 1973). Platform urbanism can offer tools to streamline and enhance that process, as discussed in the following paragraphs. However, before investigating these aspects, it is paramount to contextualize the notions of formality and informality in cities.

The formal-informal spectrum

The informal urban fabric is typically considered as a built environment designed and constructed without authorization from the state and is thought of as a spontaneous unregulated urbanization (Lejano & Del Bianco, 2018). Exemplified in slums, *favelas*, *comunas*, *kampungs*, *bidonvilles*, *barriadas*, refugee camps, and migrant settlements, the informal environment is a mechanism for the poor and disenfranchised populations to find lodging in cities that would otherwise be unaffordable. Informality is typically considered restricted to informal settlements in the city, thus considered as extraneous to the 'formal' city. This dichotomy is evident in the land-use maps generated for urban planning, where informal settlements are typically demarcated as a separate fabric with distinct boundaries that distinguish it from the city proper. However, ample evidence suggests the contrary: informality is not necessarily restricted to what is classified as 'informal settlements', and their contributions are not extraneous to the city. Dharavi, an established slum in the outskirts of Mumbai, hosts enterprises and factories allowed to operate by the state due to their economic return, not only benefiting Dharavi, but the entire city of Mumbai, despite the facilities' non-adherence to governmental codes and regulations (Sharma, 2000). The Dharavi-Mumbai relationship suggests that the informal built environment is integral to – not separate from – the city proper.

To further develop this notion, Roy and Alsayyad posit that informality in governance, economy, and processes of building the environment are not strictly an issue of the poor, but also well within the scope of the middle class and even in the elite (Roy & AlSayyad, 2004). The relationship between entrepreneurship and informality is a growing area of research, where the entrepreneur does not necessarily choose informal work out of necessity but may choose it for its profitability (Gurtoo & Williams, 2009; Maloney, 2004). This broadens the context of informality and informal work outside the built confines of informal settlement. 'Informality' then becomes a legitimate pillar of the national economy, regardless of social or economic class and regardless of location (Banks et al., 2020; Jaffe & Koster, 2019; Williams, 2007). Governmental investment in growing the informal employment sector presents a viable path toward recovery from economic hardship, especially in countries that struggle with perpetual crises or governmental inefficiencies (Debrah, 2007). In essence, informality in its spatial, economic, and societal form is inextricable from the region, the city, or the town where it dwells and must be considered as an integral facet of the region, the city, or the town where it is manifested.

In step with breaking informal labor out of its conceptual confines, informal and vernacular built environments can also be positioned as integral aspects of the town, city, or country. In viewing the vernacular, informal, and formal built environments, the boundaries sequestering the 'formal' from the 'vernacular' are not as stark as those separating the 'formal' and the 'informal'.

'Vernacular' built environment is the habitat that is constructed without the intervention of a trained architect or a specialist designer (Glassie, 1990; Oliver, 2007). It retains some of the urban and architectural historic heritage while aspiring to modernity and negotiating with the technological and resource opportunities and limitations imposed by the environment (Bourdier & AlSayyad, 1989). There is a considerable overlap – conceptual and methodological – between 'traditional' and 'vernacular' built environments, where the 'vernacular' built environment is considered a subset of, interchangeable with, or separate altogether from the 'traditional' environment (AlSayyad, 2014). Distinctions between the two can shed light on the historic and sociocultural context that produced these built environments (Carlos et al., 2015;

Kamalipour & Zaroudi, 2014) and how they reflect changes in the identity of their dwellers' communities (Jani, 2011).

The agency of the dweller in the production of the built environment creates the basis for common grounds between 'vernacular' and 'informal' urbanization. In both cases, the dweller is also often the builder who adopts community-based or traditional construction techniques to create a habitat that suits their needs, negotiates with material/economic/environmental limitations and constraints, and aspires to modernity – be it architectural or technological (Hubka, 1979; Turner & Fichter, 1972). The intermingling of vernacular and informal can be observed in the shape and distribution of public spaces (Khan et al., 2021), emerging transportation networks and the built environment's response to those networks (Samaha & Mohtar, 2020), and impact on the fabric of the built environment (Chai & Seto, 2019).

Governance challenges or absence in informal urban settlements is often viewed as a barrier to effective smart city planning. However, there is a compelling argument for reevaluating the significance of informality, shifting from perceiving it solely as a disadvantage to recognizing its potential as an advantageous space. Dweller-led development – predominant in both informal and vernacular urban milieus – challenges conventional distinctions between 'formal' and 'informal' infrastructure in policy and planning. This underscores the potential for emphasizing co-production approaches in the formulation and implementation of frameworks for resilience and smart development (Kleinhans et al., 2022; Parida, 2022). This forges new opportunities for citizen autonomy and engagement with the state and positions the informal city as a locus for continual innovation and learning, albeit following a different regimen from areas considered more 'formal' (Bobbins et al., 2024).

Platform capitalism and urbanism further challenges the formal/informal dichotomy by facilitating the transfer of aspects of informality into 'formal' systems. For example, temporary or freelance employment is prevalent in informal urban dwellings and, as mentioned earlier, benefits both the formal and the informal cities. The proliferation of platform urbanism facilitated the growth of 'gig' economies in both formal and informal urban systems (Bertolini et al., 2023, 13–45, Gig Economy, 2021). This in turn makes an 'informal' labor issue at the center of labor relationships and pushes for a re-envisioning of standardized employment relationships (Ng'weno & Porteous, 2018; Vaclacik et al., 2022). As platform urbanism takes a bigger role in community engagement especially in bottom-up planning, approaches such as DIY urbanism – a process where citizens intervene upon the urban environment without permission from the government – begin permeating the 'formal' built environments, sometimes gaining government endorsement (Foster, 2020; Sawhney et al., 2015).

The exploration of terminologies in this negotiation implies that the connection between the 'formal' and 'informal' constructs may not be a clear-cut dichotomy but rather a continuum of systems, formal and informal, built and socio-economic. Advancements in platform urbanism not only can illuminate these interconnections, but they also propose a shift in the way we conceive city development, with concepts pertaining to informality at its center. It is important to note that advocating for platforming is not a standalone solution. On the contrary, platform capitalism can potentially co-opt technological networks to serve extractive values. This co-optation may sustain or worsen existing inequities and perpetuate top-down approaches in planning (Attoh et al., 2019; Boland et al., 2022a). These pitfalls, although significant, do not deter considering platforms as tools for participatory planning and the production of local knowledge.

Platform urbanism: A conduit for informal smart city planning?

The next two subsections propose two angles to integrate platforms in smart city planning by situating the platform's intent, use, and insight offered. The first subsection highlights how investigating and contextualizing the intent of two applications (Pink Taxi and ESP) can be the foundation for formulating a city-specific smart prioritization goal. The second subsection proposes a comparative approach inspired by post-phenomenological thought (Ash & Simpson, 2016) in decoding delivery application interfaces to daylight access to food and essential services in the city.

Bridging infrastructural gaps through community-driven applications

Digital participatory planning applications directly attempt to establish transparent governance and bolster the inclusion aspect in planning. Running the gamut from data collection, scenario-based planning, and digital twinning, platforms can offer many services necessary for smart planning. The efficacy of these platforms in the planning process remains questionable, especially when the public digital literacy is not sufficiently developed to grasp the opportunities and, more importantly, limitations of the platforms in question (Boland et al., 2022b; Kitchin et al., 2021).

Indeed, digital participatory planning applications like UrbanistAI or ESRI products may encounter interruptions due to lack of public expertise, access restrictions, or simply platform malfunctions. However, such constraints do not uniformly extend to applications meditating day-to-day activities, which are normally accessible to the public to serve immediate and routine activities such as getting somewhere or buying something. This is exemplified by platforms like Uber, Amazon, and food delivery applications. The nuanced nature of limitations within participatory planning tools does not necessarily implicate the broader spectrum of digitally mediated daily applications.

The success of the latter in achieving their goals is not necessarily the sole indicator of their value. The intent of the application, especially if it is developed using local knowledge or as a market response to a local need, can formulate planning questions and priorities that very much align with smart city goals.

For instance, Pink Taxi Egypt (Pink Taxi), an application exceeding 50,000 downloads from the Google Play Store, epitomizes an endeavor by a local entrepreneur to furnish a secure transportation alternative for women in Egypt. The platform pledges the provision of female drivers for a secure and comfortable commuting experience. This initiative is not isolated. Analogous applications, such as Pink Driver and Pink Cab, have been developed to extend the same services to women and children in other regions. Pink Taxi as an application is not perfect, as evidenced by its 3.4 Google Play rating. An examination of user feedback on platforms like Google Play and Apple stores delineates sporadic operational glitches and instances of service dissatisfaction. Nevertheless, its substantial user base underscores the inadequacy of safety in public transportation, particularly for women. This highlights a potential smart city planning objective: prioritizing transportation safety, especially for women and children.

Another application is EskomSePush (ESP), an application conceived to monitor and communicate electrical-power service availability in South Africa to the public. Due to governmental, infrastructural, and financial shortcomings, the national electricity provider, Eskom, cannot provide continuous power supply. The application serves to apprise users of premeditated power outages, commonly referred to as load shedding. ESP extracts its information from Eskom's official website and supplements it with an in-app function – AskMyStreet – affording users the ability to solicit real-time power status updates from fellow residents. While the application does not purport to resolve the prevailing energy crisis, its adoption rate, exceeding 5 million downloads, substantiates its significance. It exemplifies how platform urbanism can provide solutions to systemic issues. By informing users of outages in advance, ESP minimizes disruptions to daily life. Furthermore, it influences sociocultural behaviors, as users become accustomed to planning around energy constraints and collectively sharing information. Like PinkTaxi, ESP is not the only application on the market. Similar applications, including Load Shedding Notifier and Load Shedding Schedule, corroborate the need for information dissemination to navigate daily life within infrastructural constraints. ESP has inadvertently fostered a participatory dynamic wherein citizens wield agency in defining resilience objectives, and it has set the foundation for public involvement in infrastructural planning. The reliability and public legitimacy of the platform has led to repurposing it as a platform for commercial and social activities, expanding its role beyond its initial design parameters (Hassan, 2023).

Comparing locally developed delivery applications

The concept of food delivery is not a recent innovation; it has long been a vital element of urban living. Milk delivery to households predates the car and demonstrates an extraordinary logistical effort to deliver food at an optimal temperature (Otter, 2023). A similar effort, but at a city scale, is the *Dabbawalas* – a community-driven lunchbox delivery system – exemplifies versatility, astonishing time management, and reliability in Mumbai (Sengar, 2018). Greengrocery push carts are often placed in the same category as food stalls, but notable is their mobility. They often followed a steady route, where households would expect them at a particular time in the day. Before the advent of smartphones, restaurant delivery services were already in operation but not necessarily as prevalent as today. The contemporary surge in food delivery applications, such as Postmates, DoorDash, and Grubhub, is but a natural progression. Major fast-food chains have established their own delivery infrastructure, complete with dedicated delivery personnel and user-friendly smartphone applications. Ridesharing giants like Lyft and Uber have also diversified their services to include food delivery.

The prevalence of food delivery applications allows for their comparative analysis. Variations in the design and categorization of deliverable goods offer insights into urban day-to-day life, especially when the delivery app is specific to a particular city. Recognizing the distinct challenges of informal or vernacular built environments and the diverse needs of consumers, local start-ups have tailored their delivery apps accordingly. For instance, the inclusion of a 'special notes' box has become crucial for effective wayfinding, particularly in informal built environments. Users now provide vital descriptions such as '6th floor of the building with the cellphone store on the ground level' or 'the building behind the house with the blue shutters'. Thoughtful categorization of goods and services for delivery not only enhances the user experience but also sheds light on consumer behavior.

Methodology

What follows is an exploratory comparative analysis approach to explore the relationship between food delivery platforms and urban dynamics in Amman (Jordan), Beirut (Lebanon), Mumbai (India), and Chennai (India). The methodology examines how the design and categorization of delivery platforms reflect urban contexts. The research focuses on four food delivery platforms: Swiggy, Toters, Zepto, and Talabat. The platforms were chosen for their substantial download rates (indicating prominence in their region), and their origins in local or regional start-up ecosystems (indicating likelihood of integrated local knowledge and urban features).

The comparative framework deployed to assess these applications is structured around these key categories: interface design, goods and service categorization, wayfinding and local adaptation, and custom services.

1 Interface design: The design of the platform's graphic user interface (GUI) was assessed to understand the application's facilitation of user engagement, including ease of navigation, and the overall functionality of the platform. All applications were relatively intuitive and easy to use and had very similar interfaces.
2 Goods and service categorization: The categorization of food items and other groceries, such as personal hygiene products, household essentials and electronics, was compared across platforms to identify similarities and city-specific adaptations.
3 Wayfinding and local adaptation: Special features on the platforms, such as 'special notes' for delivery instructions, were considered for their role in facilitating navigation within informal or vernacular environments. The ability of platforms to adapt to local infrastructural realities – such as informal building layouts and inconsistent street naming conventions – was a key focus of the analysis.
4 Custom services: The platforms' ability to go beyond food delivery and offer additional services, such as car cleaning, beauty treatments, or laundry, was considered to understand how platforms evolved to respond to market trends and consumer needs.

Figure 8.1 serves as a comparative diagram depicting the primary categories visible on the landing page and under the groceries section of four applications. The diagram does not include functional options (e.g., adding funds to the account or order history) and highlighted stores and trends, both illustrated in Figure 8.2. This comparative analysis is exploratory and is not meant to reach conclusive findings. Instead, its aim is to initiate an approach to understand how interface design choices in platform urbanism might yield insights into smart planning. Focusing on Swiggy and Toters, a distinctive feature is a customizable delivery option labeled as 'Genie' in Swiggy and 'Butler' in Toters. This option allows users to deliver various items within the service area, including returnable packages or items left at home, as mentioned in the Toters application. Furthermore, both Swiggy and Toters transcend the realm of mere delivery. On the one hand, Swiggy has enabled restaurant browsing and reservations in select establishments, offering an alternative to Google. Toters, on the other hand, facilitates scheduling services such as car cleaning, beauty salon visits (under self-care and health options), laundry, and cleaning services. Figure 8.2 is populated with the application verbiage, which reflects vernacular language. For example, 'Gas' refers to butane cylinders, the primary fuel source for cooking in Lebanon.

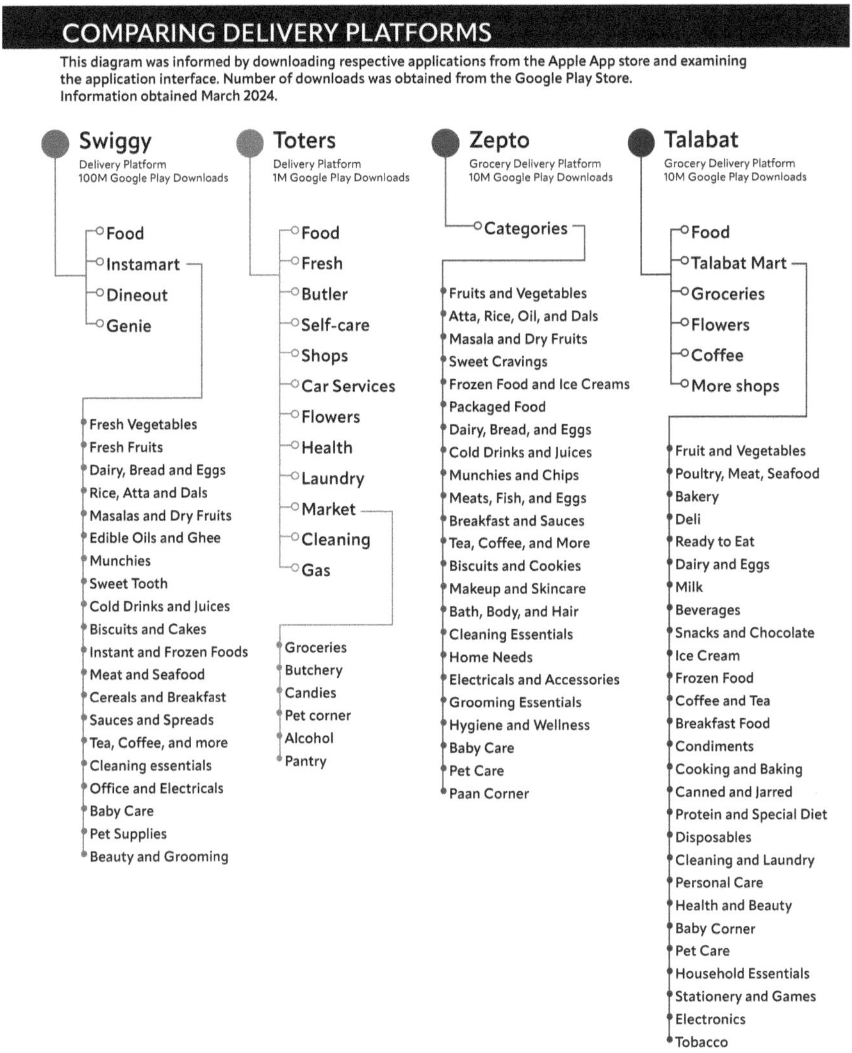

FIGURE 8.1 Comparing categorizations of delivery platforms. Prepared by Author.

Examining the extensive grocery options available on Swiggy, Zepto, and Talabat, there are a few notable observations. Categories can be overlapping or redundant: Zepto features three distinct classes for personal hygiene compared to a single category on the other platforms. This divergence might be influenced by user behavior or indicate potential design refinements in subsequent updates. Furthermore, the separate categorization of ice-cream and cold juice, particularly on Swiggy and Zepto, indicates logistical priorities for temperature-sensitive products. Interestingly, both platforms list rice, atta,

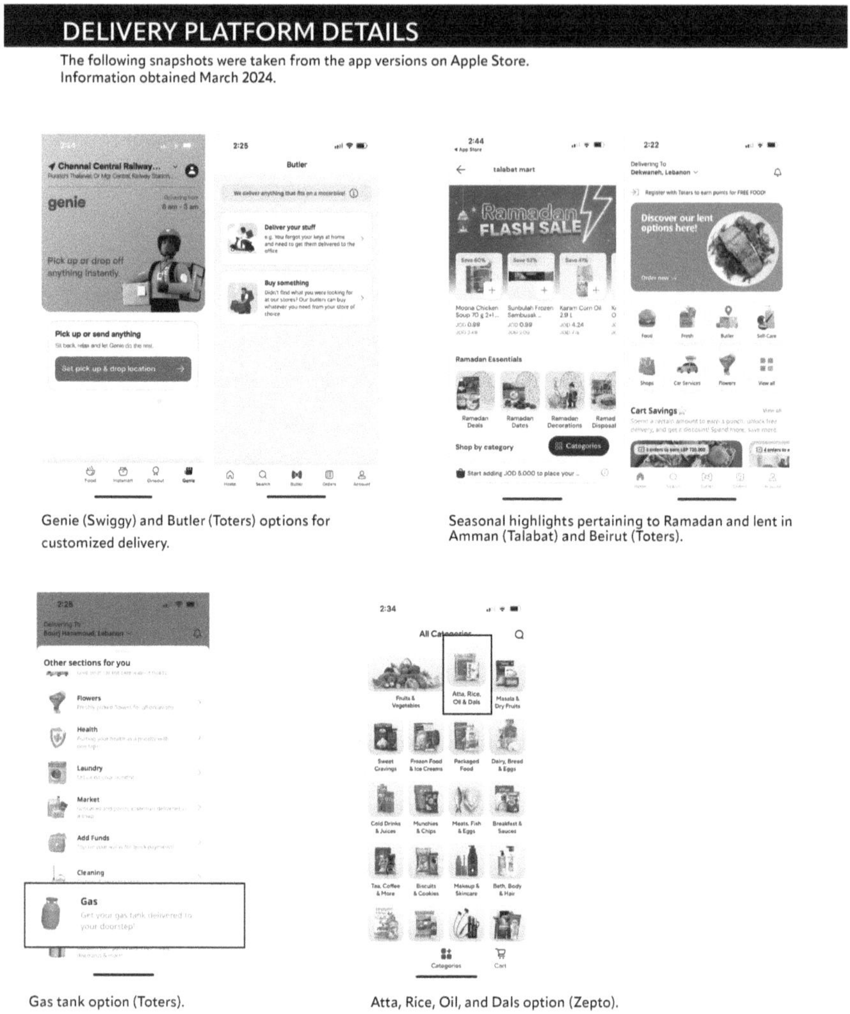

FIGURE 8.2 Delivery platform details. Screenshots collected by author from the application, downloaded from the Apple App store. Version: March 2024.

and dals as an individual category. This may be to mimic current sourcing and usage patterns and may even reflect regional differences in cuisine. In contrast, Toters presents a shorter list of options sourced featuring smaller shops and convenience stores but not larger grocery establishments. These observations underscore the dynamic interplay between user behavior, platform design, and urban contexts in shaping the offerings of food delivery applications.

The details of a platform, such as the classification of a food delivery service, can serve as a starting point for conceptualizing priorities related to walkability and access to specific goods and services within a defined service area or radius. Insight gained from the platform can serve as the groundwork for demarcating planning units such as neighborhoods or city blocks. That same insight can inform decisions related to physical infrastructure, wayfinding enhancements, and regulatory frameworks aimed at incrementally integrating smart planning principles into both formal and informal urban landscapes.

The intricacies of a platform's GUI stem from diverse sources and iterative design processes, all intended to capture and maintain user engagement. While these interactions may range from simple buyer-seller exchanges to more complex processes involving value creation and data capture, the architecture of platform ecosystems inherently catalyzes shifts in urban governance. The crucial caveat lies in ensuring that citizens maintain sovereignty over the data generated in these processes (Barns, 2019). Approaching this from a smart city planning perspective allows for the alignment of urban infrastructure goals with the inherent functions of the platform. Moreover, it provides an opportunity to conceptualize bottom-up, scale-appropriate interventions.

Discussion

Delivery platforms, particularly in densely urbanized cities, offer a lens through which to understand the dynamics of the everyday urban life. Elements within the application's design mirror the complexities of formal and informal urban infrastructure. Swiggy, Toters, Zepto, and Talabat are similarly designed and aim to offer the same range of services. Distinctions between the applications, namely the categorization of goods and services, custom services, and their support of wayfinding fields, can be indicative of localized urban needs. The inclusion of a 'special notes' allows users to enter specific wayfinding instructions that accommodate informal layouts of cities. On the opposite end, this also fosters standard practices among delivery staff. This feature underscores the importance of adapting digital interfaces to the specific needs of users navigating urban landscapes. The introduction of additional services, such as Swiggy's 'Genie' and Toters' 'Butler', highlights how delivery platforms can potentially mitigate gaps in local infrastructure. These platforms become instruments not just for commerce but also for managing everyday life in cities. From an urban planning perspective, the categorization of goods on these platforms offers insights into consumer preferences and the transportation infrastructure.

In a greater context, platform urbanism has an undeniable role in shaping future cities tailored to the needs of its citizens. However, there are many limitations or aspects that could hinder or even reverse the positive role of platforms. The most evident aspect is data sovereignty and privacy, and how platforms, especially those with a large user base, will likely perpetuate or worsen the power imbalance between users and platform owners/developers (Attoh et al., 2019; Boland et al., 2022a). This is especially relevant in platform urbanism in the service of smart city planning, as user input data are essential to gain insight into the specificities of urban dynamics. Data mining could facilitate manipulating users (citizens) into behaviors and purchases that do not serve their interests, but rather the interests of the tech industry and companies benefiting from the harvested data (Zuboff, 2019). The failure of governments to provide adequate infrastructural services in cities often extends to their inability to regulate platform spaces and company practices effectively. While platforms like Android and Apple app stores impose certain requirements on applications, offering some level of user protection, these measures are far from comprehensive. Platform standards and regulations often operate beyond the scope of local governance, and the absence of robust accountability mechanisms further exacerbates the vulnerability of users.

Conclusion: Mapping through platform urbanism

With 'smart cities' circulating the global sphere for the last decade, scholarly efforts are attempting to define what a smart city is, as the definition continues to shift and change depending on the context. What is certain is that it is a desirable effort for digitally mediating an improved, inclusive, and futuristic city. These efforts are normally considered in cities where the infrastructure is already viable and there is buy-in from local and national governance. Urban milieus perceived outside the formal constraints of planning are not considered for smart city approaches. Critiques of smart city planning question the 'presentism' of these approaches and call for a wider temporal lens to understand the current urban state vis-à-vis its past, and an honest assessment of the viability of these solutions (Datta, 2019; Srnicek, 2017). This wider spatio-temporal lens would facilitate the inclusivity of urban fabrics in their varying degrees of formality in smart city planning efforts.

While fragmented or absent governance and failing infrastructure constitute major roadblocks, the chapter argues that local knowledge can sidestep existing failures and propose a different path toward smart cities. In application, platforms that mediate the citizen's daily interaction with the city can become useful. The queries and services offered by the most successful applications shed light on how the city operates and how it could be brought into the smart city movement.

Tools to investigate the platform itself from the user's point of view are of equal importance. In platform urbanism, the smartphone becomes an instrument for generating knowledge and the production of the city. Investigating the platform's intent, how the intent is articulated through its interface, and how it is eventually used, begins to plot how platforms shape and guide action without tethering every strain of analysis to a presumption of manipulation (Ash et al., 2018). Platforms can demonstrate community approaches to overcome infrastructural shortcomings, such as in the case of South Africa's ESP and Egypt's Pink Taxi. Platforms can also begin to construct economic ecosystems that could be gleaned through delivery services.

This exploratory chapter compares four delivery applications, focusing on specificities of the user interface. This approach has its limitations, and there is a need for analytical methods based on post-phenomenology to synthesize information from platforms that can encompass aspects such as context, functionality, and data infrastructure. This includes considerations for platform infrastructure, the human-technology boundary (Matthews, 2021), and open-source approaches (Bradley, 2015). Platform investigation can be furthered into back-end components, including funding and financial structures, the relationship between developers, government, and users, as well as issues related to data sovereignty.

In conclusion, this chapter underscores the potential of locally developed platform infrastructure in bringing smart planning to informal urban environments. By analyzing the intricacies of food delivery and participatory planning applications, the study demonstrates how these platforms can offer insights into urban dynamics and priorities. Starting with framing the notion of the formal-informal as a spectrum allows for advancements in platform urbanism to illuminate interconnected systems. A broader perspective on smart city planning that includes informal cities encourages co-production approaches for resilience and innovation. While the prospects for platform urbanism are promising, the outcome is not guaranteed. Indeed, the existing infrastructure can easily perpetuate extractive processes, harvesting data and stripping the public from data sovereignty. Overall, the findings advocate for a nuanced understanding of urban development, acknowledging the agency of dwellers in shaping their digital and physical environments.

References

Abdi, H., & Shahbazitabar, M. (2020). Smart city: A review on concepts, definitions, standards, experiments, and challenges. *Research Article Journal of Energy Management and Technology (JEMT)*, *4*, 1. https://doi.org/10.22109/jemt.2020.206444.1205

Albino, V., Berardi, U., & Dangelico, R. M. (2015). Smart cities: Definitions, dimensions, performance, and initiatives. *Journal of Urban Technology* 22 (1), 3. doi: 10.1080/10630732.2014.942092

AlSayyad, N. (2014). *Traditions: The "real", the hyper, and the virtual in the built environment.* Routledge/Taylor & Francis Group.

Ash, J., Anderson, B., Gordon, R., & Langley, P. (2018). Unit, vibration, tone: a post-phenomenological method for researching digital interfaces. *Cultural Geographies*, *25*(1), 165–181. https://doi.org/10.1177/1474474017726556
Ash, J., & Simpson, P. (2016). Geography and post-phenomenology. *Progress in Human Geography*, *40*(1), 48–66. https://doi.org/10.1177/0309132514544806
Attoh, K., Wells, K., & Cullen, D. (2019). "We're building their data": Labor, alienation, and idiocy in the smart city. *Environment and Planning D: Society and Space*, *37*(6), 1007–1024. https://doi.org/10.1177/0263775819856626
Banks, N., Lombard, M., & Mitlin, D. (2020). Urban informality as a site of critical analysis. *Journal of Development Studies*, *56*(2), 223–238. https://doi.org/10.1080/00220388.2019.1577384
Barns, S. (2019). *Platform urbanism: Negotiating platform ecosystems in connected cities*. Springer Nature.
Bertolini, A., Graham, M., Neerukonda, M., Ojanperä, S., Parthasarathy, B., Srinivasan, J., Taduri, P., & Ustek-Spilda, F. (2023). Platformizing informality, one gig at a time. In A. Surie & U. Huws (Eds.), *Platformization and Informality : Pathways of change, alteration, and transformation* (pp. 13–45). Springer International Publishing. https://doi.org/10.1007/978-3-031-11462-5_2
Bobbins, K., Caprotti, F., de Groot, J., Pailman, W., Moorlach, M., Schloemann, H., Densmore, A., Finlay, K., Fischat, E., Siwali, S., & Links, J. (2024). Smart and disruptive infrastructures: Re-building knowledge on the informal city. *Urban Studies*, *61*(1), 165–179). https://doi.org/10.1177/00420980231172582
Boland, P., Durrant, A., McHenry, J., McKay, S., & Wilson, A. (2022a). A 'planning revolution' or an 'attack on planning' in England: digitization, digitalization, and democratization. *International Planning Studies*, *27*(2), 155–172. https://doi.org/10.1080/13563475.2021.1979942
Boland, P., Durrant, A., McHenry, J., McKay, S., & Wilson, A. (2022b). A 'planning revolution' or an 'attack on planning' in England: digitization, digitalization, and democratization. *International Planning Studies*, *27*(2), 155–172. https://doi.org/10.1080/13563475.2021.1979942
Bourdier, J. P., & AlSayyad, N. (1989). *Dwellings, settlements, and tradition: Cross-cultural perspectives* (International Association for the Study of Traditional Environments, Ed.). University Press of America.
Bradley, K. (2015). Open-source urbanism: Creating, multiplying and managing urban commons. *Footprint: Delft Architecture Theory Journal*, *16*. https://doi.org/doi.org/10.7480/footprint.9.1.901
Caprotti, F., Chang, I.-C. C., & Joss, S. (2022). Beyond the smart city: A typology of platform urbanism. *Urban Transformations*, *4*(1). https://doi.org/10.1186/s42854-022-00033-9
Carlos, G. D., Correia, M., & Rocha, S. (2015). Vernacular architecture? In M. Correia, P. Lourenco, & H. Varum (Eds.), *Seismic retrofitting: Learning from vernacular architecture*. CRC Press.
Chai, B., & Seto, K. C. (2019). Conceptualizing and characterizing micro-urbanization: A new perspective applied to Africa. *Landscape and Urban Planning*, *190*(June), 103595. https://doi.org/10.1016/j.landurbplan.2019.103595
Datta, A. (2019). Postcolonial urban futures: Imagining and governing India's smart urban age. *Environment and Planning D: Society and Space*, *37*(3), 393–410. https://doi.org/10.1177/0263775818800721
Debrah, Y. A. (2007). Promoting the informal sector as a source of gainful employment in developing countries: Insights from Ghana. *International Journal of*

Human Resource Management, *18*(6), 1063–1084. https://doi.org/10.1080/09585190701321716

Foster, N. (2020). Rethinking the right to the city: DIY urbanism and postcapitalist possibilities. *Rethinking Marxism*, *32*(3), 310–329. https://doi.org/10.1080/08935696.2020.1780668

Gig economy. (2021, November 11). Gig workers have derailed the formal-informal divide. *The Hindu Businessline*.

Glassie, H. (1990). Architects, vernacular traditions, and society. *Traditional Dwellings and Settlements Review*, *1*(2), 9–21.

Gracias, J. S., Parnell, G. S., Specking, E., Pohl, E. A., & Buchanan, R. (2023). Smart cities—A structured literature review. *Smart Cities*, 6(4), 1719–1743. https://doi.org/10.3390/smartcities6040080

Gurtoo, A., & Williams, C. C. (2009). Entrepreneurship and the informal sector. *The International Journal of Entrepreneurship and Innovation*, *10*(1), 55–62. https://doi.org/10.5367/000000009787414280

Hassan, S. (2023, June 12). South Africans are using a power outage notification app to sell weed, find odd jobs. *Rest of World*.

https://smartcities.gov.in/. (n.d.). *Smart City Mission*.

https://www.neom.com/en-us. (n.d.). *Neom*.

Hubka, T. (1979). Just folks designing vernacular designers and the generation of form. *Journal of Architectural Education*, *32*(3), 27–29. https://doi.org/10.1080/10464883.1979.10758609

Jaffe, R., & Koster, M. (2019). The Myth of formality in the global north: Informality-as-innovation in dutch governance. *International Journal of Urban and Regional Research*, *43*(3), 563–568. https://doi.org/10.1111/1468-2427.12706

Jani, V. (2011). *Diversity in design: Perspectives from the Non-Western World*. Bloomsbury Academic.

Kamalipour, H., & Zaroudi, M. (2014). Sociocultural context and vernacular housing morphology: A case study. *Current Urban Studies*, *02*(03), 220–232. https://doi.org/10.4236/cus.2014.23022

Khan, F. M., Pafka, E., & Dovey, K. (2021). Understanding informal functional mix: morphogenic mapping of Old Dhaka. *Journal of Urbanism: International Research on Placemaking and Urban Sustainability*, *00*(00), 1–19. https://doi.org/10.1080/17549175.2021.1979085

Kitchin, R., Young, G. W., & Dawkins, O. (2021). Planning and 3D spatial media: progress, prospects, and the knowledge and experiences of local government planners in Ireland. *Planning Theory and Practice*, 22(3), 349–367. https://doi.org/10.1080/14649357.2021.1921832

Kleinhans, R., Falco, E., & Babelon, I. (2022). Conditions for networked co-production through digital participatory platforms in urban planning. *European Planning Studies*, *30*(4), 769–788. https://doi.org/10.1080/09654313.2021.1998387

Kourtit, Karima and Peter Nijkamp. 2012. Smart cities in the innovation age. *Innovation: The European Journal of Social Science Research*, *25* (2), 93. doi:10.1080/13511610.2012.660331

Lejano, R. P., & Del Bianco, C. (2018). The logic of informality: Pattern and process in a São Paulo Favela. *Geoforum*, *91*(February), 195–205. https://doi.org/10.1016/j.geoforum.2018.03.005

Maloney, W. F. (2004). Informality revisited. *World Development*, *32*(7), 1159–1178. https://doi.org/10.1016/j.worlddev.2004.01.008

Matthews, A. (2021). Blurring boundaries between humans and technology: postdigital, postphenomenology and actor-network theory in qualitative research. *Qualitative Research in Sport, Exercise and Health, 13*(1), 26–40. https://doi.org/10.1080/2159676X.2020.1836508

Ng'weno, A., & Porteous, D. (2018). *The Informal Sector and the Gig Economy in AFRICA Let's Be Real: The Informal Sector and the Gig Economy are the Future, and the Present, of Work in Africa.*

Oliver, P. (2007). Built to meet needs: Cultural issues in vernacular architecture. In *Built to meet needs: Cultural Issues in vernacular architecture.* https://doi.org/10.4324/9780080476308

Otter, C. (2023). Milk in motion: Logistical geographies in twentieth-century Britain. *Global Food History, 9*(1), 47–71. https://doi.org/10.1080/20549547.2022.2138068

Parida, D. (2022). Fantasy visions, informal urbanization, and local conflict: An evolutionary perspective on smart city governance in India. *GeoJournal, 87*(6), 4707–4718. https://doi.org/10.1007/s10708-021-10521-3

Pink taxi. https://pinktaxi.net/

Roy, A., & AlSayyad, N. (2004). *Urban informality: transnational perspectives from the Middle East, Latin America, and South Asia*, edited by Ananya Roy and Nezar AlSayyad. Md Lanham. Lexington Books .

Roy, S. (2016). The Smart City Paradigm in India: Issues and challenges of sustainability and inclusiveness. *Social Scientist, 44*(5/6), 29–48. http://www.jstor.org/stable/24890283

Samaha, P., & Mohtar, A. (2020). Decoding an urban myth: An inquiry into the Van line 4 system in Beirut, Lebanon. *Journal of Transport Geography, 85*(August 2019), 102743. https://doi.org/10.1016/j.jtrangeo.2020.102743

Sawhney, N., de Klerk, C., & Malhotra, S. (2015). Civic engagement through DIY urbanism and collective networked action. *Planning Practice & Research, 30*(3), 337–354. https://doi.org/10.1080/02697459.2015.1054662

Sengar, R. (2018, December 1). The story of Mumbai's Dabbawalas. *Times of India.*

Sharma, K. (2000). *Rediscovering Dharavi: Stories from Asia's largest slum.* Penguin Books.

Smith, R. W. (1973). A theoretical basis for participatory planning. *Policy Sciences, 4*(3), 275–295. https://doi.org/10.1007/BF01435125

Srnicek, N. (2017). *Platform capitalism.* Polity.

Turner, J. F. C., & Fichter, R. (1972). *Freedom to build: Dweller control of the housing process.* Macmillan. https://books.google.com/books?id=lwaxAAAAIAAJ

Vaclacik, M. C., Oltamari, A. P., & De Oliveira, S. R. (2022). Managing the informal economy: A theoretical debate in the light of gig economy. *Cadernos EBAPE.BR, 20*(2), 247–258. https://doi.org/10.1590/1679-395120210065x

Williams, C. C. (2007). The nature of entrepreneurship in the informal sector: Evidence from England. *Journal of Developmental Entrepreneurship, 12*(02), 239–254. https://doi.org/10.1142/s1084946707000654

Zuboff, Shoshana. 2019. *The age of surveillance capitalism: The fight for a human future at the new frontier of power.* New York: PublicAffairs.

PART III

Smart cities and the law

9

SMART CITIES AS NEW BATTLEGROUNDS IN THE GEOPOLITICS OF INTERNATIONAL LAW

Shisong Jiang and Xinyue Zheng

Introduction

The concept of smart cities has gained momentum over the past decade, driven by rapid urbanization and technological advancements (Albino et al., 2015). These cities harness a plethora of technologies, including artificial intelligence (AI), Internet of Things (IoT), big data analytics, cloud computing, and advanced telecommunication networks like 5G (Javed et al., 2022). At their core, smart cities aim to optimize city functions, enhance sustainability, and improve the quality of life for residents through data-driven decision-making and efficient resource management (Lim et al., 2023). They integrate various systems – transportation, healthcare, energy, water management, waste disposal, and security – into a cohesive and responsive network (Ismagilova et al., 2019). IoT devices play a critical role, with sensors and cameras collecting real-time data, which is then processed and analyzed to improve city services and infrastructure (Samih, 2019). AI and machine learning algorithms assist in predicting trends, managing resources, and providing personalized services to residents (Ullah et al., 2020).

The evolution of smart cities is increasingly entangled in the broader geopolitical contest (e.g., Datta & Odendaal, 2019; Faraji et al., 2023; Rowland et al., 2022), particularly between the United States (US) and China (particularly see Weber, 2023). This competition extends beyond the race for technological innovation; it encapsulates differing ideologies and visions for the future of urban governance. For instance, the US approach to smart cities often emphasizes enhancing quality of life, economic efficiency, and citizen engagement, reflecting a more market-driven and decentralized model of development. In contrast, China's approach is notably different, focusing on

DOI: 10.4324/9781003439325-13

surveillance, social control, and centralized management, underpinned by its authoritarian governance model (Ekman, 2019). Chinese firms, sometimes backed by the state, have been actively exporting smart city technologies, particularly to countries in Asia, Africa, and the Middle East, intertwining economic aid, digital infrastructure development, and political influence (Weber, 2023).

Smart cities introduce complex new dynamics at the intersection of technology, governance, and geopolitics, presenting unique challenges to the established norms of (international) law (e.g., Brown, 2019; He et al., 2022; Losavio et al., 2018; Ranchordás & Klop, 2018). These challenges arise from issues related to data sovereignty, privacy, surveillance, and the digital divide. The deployment of smart city technologies can lead to mass data collection and surveillance, raising significant privacy concerns and potential human rights violations, particularly in authoritarian regimes (despite the controversy, see, e.g., Tsai et al., 2023; Vinha, 2023). Additionally, the global reach of smart city technologies, often controlled by a few dominant players, raises questions about data sovereignty and the influence of foreign entities on domestic governance (e.g., March & Ribera-Fumaz, 2018; Momen, 2023).

While smart city development is a global phenomenon, with significant initiatives in Europe, Asia, and other regions, this chapter focuses primarily on China and the US for several reasons. First, these two nations represent the world's largest economies and are at the forefront of technological innovation (Bustamante & Calidas, 2024; Cugurullo, 2020). Second, they embody contrasting models of smart city development – China's state-driven, surveillance-oriented approach versus the US's market-led, citizen-centric model – which provides a clear illustration of how different political systems shape smart city implementation (Ekman, 2019). Lastly, the geopolitical rivalry between China and the US in the technological sphere makes their smart city initiatives particularly relevant to the discussion of international law and global governance (Swiney, 2020). While examples from Europe and other regions are mentioned where relevant, the China-US dynamic serves as the primary case study to explore the broader implications of smart cities on the geopolitics of international law.

Against this backdrop, this chapter examines the following research question: why and to what extent have smart cities emerged as a new battleground in the geopolitics of international law? This question is crucial as it illuminates the evolving intersection of technology, urban governance, and international law in an increasingly digitalized world. The relevance lies in understanding how smart city developments are reshaping legal frameworks, challenging existing norms, and potentially creating new paradigms in international law. In this light, the overall impact on law is multifaceted: it necessitates the development of new legal instruments to govern cross-border data flows, redefines concepts of sovereignty in the digital realm, and requires a reevaluation of human rights protections in technologically advanced urban environments. Moreover, it highlights the need for international legal systems to adapt to the rapid pace of technological change and the growing influence of non-state actors, particularly cities and tech companies, in shaping global governance structures.

It argues that major powers like China and the US view the development of smart city technologies as a strategic priority and are heavily investing in them as a means of expanding soft power and competing technologically on a global scale. For instance, China's "Smart City" initiative, part of its broader Digital Silk Road project, has invested billions in over 500 smart city pilot projects both domestically and internationally, particularly in Southeast Asia and Africa. In parallel, the US, through its Smart Cities and Communities program under the National Science Foundation, has allocated substantial funding for research and development of smart city technologies, while also promoting its vision of open, democratic smart cities through initiatives like the Global Tech Challenge (Kurlantzick & West, 2020). These investments not only showcase technological prowess but also promote each country's distinct model of urban governance and data management on the global stage. More importantly, their smart city initiatives are expanding the scope of international legal debates and contestation. At the same time, cities themselves are playing an increasingly prominent role in shaping the norms, rules, and standards that govern smart city development and operation.

This chapter will be structured in the following manner. Section 2 provides background on the concept of the "geopolitics of international law" and analyzes how cities are being used as political tools by states to compete for influence and advance national interests. It examines smart cities specifically as an emerging site of international legal contestation related to their governance. Section 3 provides a case study comparing the smart city strategies and initiatives of China and the US, with a focus on their respective geopolitical motivations. Section 4 assesses how cities themselves are influencing the norms and legal frameworks related to smart cities, as well as the constraints they face in this role. It also examines the potential for cities to employ international legal mechanisms to uphold standards and hold states accountable for commitments relating to sustainable smart city development. Finally, Section 5 summarizes the key arguments and reflects on future trajectories in the geopolitics of smart cities from the perspective of international law.

Overall, this chapter illuminates smart cities as an increasingly important terrain where the broader dynamics in the geopolitics of international law unfold. As the global race to develop smart cities accelerates, their governance has legal implications that merit further academic examination. The paper explores these dimensions in depth by situating smart cities in the context of states' use of law as an instrument to wield power and counterbalancing moves by cities to play a more active role in setting and upholding legal standards. This specific context was chosen because it provides a crucial lens through which to understand the evolving power dynamics in international law. Traditionally, states have been the primary actors in shaping international legal norms (Marks, 2006). However, the emergence of smart cities challenges this state-centric model, introducing new actors (cities and tech companies) and new domains of governance (digital infrastructure and data) (especially see Aust & Nijman, 2021). By examining how states attempt to use law to maintain their influence in this new terrain, while cities simultaneously assert their own agency, we can better understand the changing nature of international law in the digital age. This context allows us to explore how the development of smart cities is not just a technological phenomenon, but a profound shift in the landscape of global governance and legal frameworks. It culminates in a reflection on the future of global smart city development from the standpoint of international law and governance.

The geopolitics of international law and the role of cities

This section explores the historical development of the geopolitics of international law, the evolving role of cities as instruments of state power, and the emergence of smart cities as new sites of geopolitical competition. Unfolding these foundational elements will set the stage for the subsequent analysis of the China-US case study and the role of cities in shaping smart city governance.

The concept of geopolitics of international law

The "geopolitics of international law" refers to the strategic use of international law and associated legal institutions by states to advance national interests and influence global order (also see Orakhelashvili, 2008). Dominant world powers have long exploited their geopolitical and economic might to shape the development of international rules and norms in ways that augment their own power and interests (Dezalay & Garth, 2002). By sponsoring treaties, spearheading global regulatory initiatives, controlling key appointments in multilateral legal bodies, and shaping legal scholarship, hegemonic states mold international law as an "ideological armament" to consolidate their leadership over peripheries (Anghie, 2005).

While the term "geopolitics of international law" has gained greater currency recently, the underlying dynamics it encapsulates are far from new. As early as the 17th-century Treaty of Westphalia, dominant European powers crafted international legal principles of sovereignty and non-intervention that locked in their control over colonies abroad while minimizing intervention across territories within Europe (Zarakol, 2017). Throughout the 19th century, imperial Britain imposed unequal treaties backed by gunboat diplomacy across Asia and the Middle East to entrench its economic and political domination (Hulle, 2016). The US marshaled international institutions like the World Bank and the International Monetary Fund following World War II to institute the contemporary neoliberal order that maintains the global supremacy of Wall Street and Silicon Valley (Chimni, 2018; Harvey, 2005).

Understanding international law as an ideological apparatus deeply imbricated within global power networks highlights its nature more as a site of struggle between competing state interests rather than a neutral realm removed from politics (Krisch, 2021). As new technologies and threats emerge within a more multipolar 21st-century world order, dominant states continually leverage their material and institutional capabilities to shape international legal debates over nascent issues in ways that serve their own priorities and strategic visions (Mearsheimer, 2019). Cities have become critical tools within these unfolding battles over governance of emerging domains (see, e.g., Aust & Nijman, 2023; Jiang, 2020).

Cities as instruments of state power and influence

Cities have long served key roles within empires and nation-states as political capitals, military fortresses, transportation hubs, manufacturing centers, and markers of modernity and prestige. However, contemporary conditions of globalization have amplified both the symbolic and functional importance of cities as platforms through which states project influence abroad and compete for global capital and talent (Curtis, 2016). This subsection details some of the mechanisms through which states exploit city development to burnish and employ soft power in the service of foreign policy goals.

One way that dominant states utilize cities as geopolitical instruments is by funneling resources into lavish megaprojects that garner global media attention and boost nation-branding narratives (Crouch & Galès, 2012). Gulf states like the United Arab Emirates and Qatar have poured tens of billions of dollars into flamboyant cultural complexes, skyscrapers, artificial luxury islands, and event spectacles with thin economic rationale beyond raising their profiles as globally ambitious rising powers (Henderson, 2021). Turkey has explicitly framed its plans to construct an eight-lane shipping canal flanked by expansive new smart cities as consecrating its status as a regional geostrategic leader (Eldem, 2021). States bankroll such grandiose urban developments not for commercial prospects but rather soft power returns – as calling cards of sovereignty, modernity, and economic dynamism to outside constituencies (Swiney, 2020).

In addition to symbolic nation branding functions, states directly sponsor cities as communications hubs and gateways to regional markets. Special economic zones, export processing zones, and science parks established across Southeast Asia, Latin America, and Africa with government backing concentrate advanced transportation and telecom infrastructure to attract multinational corporate activity that upgrades domestic manufacturing and services provision (Papanikolaou et al., 2022). By subsidizing built environments primed for global business, states drive local employment and economic upgrades that communicate policy competence, nurture loyal political constituencies, and cement leadership over neighboring countries also vying for international commercial interest and associated knowledge transfers.

States similarly tap favored cities as diplomatic arms that amplify policy initiatives abroad (Acuto, 2013). For instance, Mexico, Brazil, and Argentina have designated particular cities as coordination points for regional bodies like the Latin American Integration Association and Mercosur alliance, hosting headquarters facilities along with regular ministerial summits (Criekemans, 2010). States authorize select cities to channel diplomatic exchanges around issues like climate change, global health, and disaster response that boost international leadership credentials (Acuto, 2013). Moreover, this extensive state sponsorship of cities as platforms to promote soft power abroad seems likely to intensify, given modern cities' economic and demographic growth. Over 50% of humanity already resides in cities that generate 80% of global GDP – concentrations of people, wealth, and delivery systems that have expanded government efforts to harness urban potential (United Nations, 2019). As more activity converges within their boundaries, strategically significant cities will play amplified roles in shaping state foreign policies far into the future.

Smart cities as new sites of geopolitical competition

The rapid development of smart city technologies over the past decade has opened up a major new arena of geopolitical competition between states at the frontier of international governance (e.g., Datta & Odendaal, 2019; Ekman, 2019; Rowland et al., 2022; Weber, 2023). Smart cities deploy vast arrays of digital sensors, real-time data analytics, automated transport and logistics networks, AI-powered public service delivery platforms, drone and satellite monitoring systems, control room nerve centers, and ubiquitous public video surveillance to enhance urban operations and responsiveness (Calzada, 2018b). They serve as urban "laboratories" to test innovative technological solutions and next-generation infrastructure systems (Kitchin et al., 2018).

However, beyond purely municipal domains, the advent of smart cities also carries major international power implications, given their roles in incubating frontier technologies with economic and military applications (Molchanov & Molchanova, 2022). Models of urban governance engineered within these test beds shape global standards over critical issues like data ownership, privacy, algorithmic transparency, and export/import controls over sensitive technologies. Already, expert market research shows that the smart cities' market is projected to grow from $581.89 billion in 2023 to $744.15 billion in 2026, with a compound annual growth rate of 8.1% (GlobalData, 2022).

For major powers like China and the US, establishing supremacy in smart city development represents a critical pathway toward technological leadership across the broader digital economy (Noor, 2023). Success in this domain promises spillover advantages in other high-tech arenas like wireless communications, autonomous vehicles, renewable energy, semiconductors, robotics, and AI – all pillars of global economic and military power over the 21st century.

The governance of smart cities as a site of international legal contestation

The unfolding global race to dominate smart city leadership has generated a major new site of international legal controversy and standard setting (e.g., Marchetti et al., 2019; Robin & Acuto, 2018). Divergent state-led models guided by contrasting political values promote alternative visions over appropriate rules, principles, and technical protocols across a host of novel policy domains with inadequate existing legal coverage (Joss et al., 2017). As Chinese smart city platforms diffuse internationally, they challenge Western data privacy frameworks, mirroring broader ideological struggles between authoritarian and liberal democratic states (Große-Bley & Kostka, 2021).

One major axis of global standard setting where smart city development has activated legal tensions involves data governance (Choenni et al., 2022). As urban operations grow increasingly "datafied" through proliferating sensors and monitoring systems, new frameworks are required over issues of data transparency, portability, privacy, storage, and ownership (Barns, 2020). However, starkly differing smart city templates promoted by China and the West institutionalize contradictory approaches. Under its "City Brain" model, China aggregates control center sensor feeds, social media data, and footage from mandatory surveillance cameras across entire urban landscapes to train automated governance algorithms (J. Zhang et al., 2019). By contrast, the European Commission has moved toward a "federated learning" structure where privacy-preserving machine learning models are trained locally before aggregated into a global model (Al-Huthaifi et al., 2023). As Chinese developers aggressively market their smart city wares with lax built-in data protections abroad, these models clash around accountability and ethics (F. Yang & Xu, 2018).

A related realm of legal dispute triggered by smart city expansion involves technology transfer and 5G network infrastructure control. As major American and European telecom equipment manufacturers like Nokia, Ericsson, and Cisco have fallen behind Huawei and ZTE over 5G, worries have grown globally over dependency upon Chinese-controlled communication networks deeply embedded within smart cities (Shi-Kupfer & Ohlberg, 2019). The politics of 5G have accordingly intensified debates at the World Trade Organization (WTO) and beyond over appropriate mechanisms to regulate cross-border data flows and limit Made in China high-tech penetration based on cybersecurity and national security rationales (Kania, 2019). US-led demands for stricter international controls over next-generation technology transfers to shield its eroding innovation edge now powerfully intersect with its smart city rivalry with China (Bateman, 2022).

Looking forward, the unfolding global race to dominate smart city development seems highly likely to amplify international legal divisions along both North-South and East-West lines (e.g., Reuter, 2020). As these urban epicenters of technological innovation expand in reach and influence, their governance design principles appear destined to shape trailblazing norms, standards, and regulatory models with widespread international consequences (Badran, 2023). Much as imperial powers anchored liberal internationalism via colonial cities and sponsorship of Westphalian statehood under the League of Nations Mandate system at the start of the 20th century (Getachew, 2019), major powers now jostle to imprint the still highly fluid realm of smart urbanization – whose contours will indelibly mold coming technology order.

Case study: Smart cities in China and the US

Having established the broader context of smart cities in the geopolitics of international law, this section focuses on China and the US as leading global powers in smart city development. While other regions, particularly Europe (Correia et al., 2022), have made significant strides in this field, the China-US dynamic offers the most stark contrast in approaches and geopolitical implications. By examining their technological strategies, governance models, and international influence efforts, we can gain valuable insights into how smart cities are becoming a critical battleground in the ongoing geopolitical and legal competition between China and the US.

China's technological approach to smart cities

As introduced previously, China has made smart city development a top strategic priority under its Digital Silk Road and Belt and Road initiatives to expand technological capacity and extend geostrategic influence across

Eurasia (Y. He, 2024; Heeks et al., 2024; Seoane & Velasco, 2024). The country's technological model for constructing intelligent urban spaces emphasizes the rapid buildout of advanced digital infrastructure like ubiquitous sensor networks, AI-powered control rooms, and cloud analytics ecosystems channeling real-time data streams into actionable intelligence for optimizing municipal service delivery (Große-Bley & Kostka, 2021).

Shenzhen exemplifies China's state-driven, technology-centric approach to smart city development by implementing an extensive network of sensors and cameras that can integrate data from various urban systems into a centralized platform (Xie et al., 2024). As with Hangzhou's "City Brain" (Caprotti & Liu, 2022), this system aggregates information from traffic cameras, environmental sensors, public transportation, and other municipal services to create a comprehensive urban management tool (R. Hu, 2019). While the exact number of data sources is not publicly confirmed, the scale is significant, allowing for real-time monitoring and management of urban flows. The system's capacity to process large amounts of data enables it to track anomalies, assist in decision-making, and provide predictive analytics for urban management (Xie et al., 2024).

Shenzhen's approach does reflect China's emphasis on intelligent command and control systems leveraging big data and AI for centrally administered predictive governance and automated regulation enforcement – an institutional priority reflected across its sweeping smart city designs (Große-Bley & Kostka, 2021). This approach aligns with China's legacy of centralized technocratic planning from Beijing downwards through provincial and municipal conduits that furnished rapid infrastructure development nationwide over recent decades (Caprotti & Liu, 2022). It extends familiar governance logic into emerging digital spaces as sensors, algorithms, and cloud analytics unlock novel possibilities for granular, dynamic, and automated oversight of civic life (Xu et al., 2024).

Critically, China's smart city model also involves a close cooperation between urban governance functions and commercial technology providers, though the exact nature of this relationship may differ from the original description (Caprotti & Liu, 2022). The Shenzhen government collaborates with high-tech enterprises through various modes of contracting and outsourcing (Xie et al., 2024). State-owned telecom carriers provide connectivity infrastructure, while locally based tech giants like Huawei and Tencent play pivotal roles in providing hardware and software solutions, respectively (Xie et al., 2024). However, it is important to note that while these companies are deeply involved in smart city projects, their access to sensitive data is strictly monitored and controlled by the government (Xie et al., 2024). This structure aims to build local technological capabilities while ensuring that the government maintains primary control over data and system management (Ju et al., 2024).

US civic emphasis for smart cities

In contrast to China's centralized approach (Q. Hu & Zheng, 2021), the US promotes a model for smart cities focused more on civic participation, inclusion, and distributed innovation across communities (Wiig, 2015; Wilson & Chakraborty, 2019). Where Beijing prioritizes sensor connectivity and computational power for centralized monitoring and response coordination firewalled across internal security services, Washington emphasizes grassroots co-design processes harnessing local creativity for solutions advancing collaborative democracy (Anderson & Jung, 2023). This alternative vision champions bottom-up digital empowerment of residents through communal tools like participatory budgeting platforms, open contracting systems, and robust transparency portals over tech-led municipal automation (e.g., Gilman, 2016; see others in Menendez-Blanco & Bjørn, 2022; Palacin et al., 2024).

For example, milestone smart cities projects like New York's LinkNYC initiative replace old payphone stands with decentralized networks of kiosks offering free public Wi-Fi and device charging stations (NYC Office of Technology and Innovation, n.d.; see more in Maier et al., 2018). The LinkNYC platform furnishes base connectivity infrastructure for residents and entrepreneurs rather than AI-directed policy interventions (Shah et al., 2019). Designed to meet future civic tech needs, it supports community engagement and local innovation (Amsellem, 2021). Similarly, cities like Louisville, San Francisco, and Detroit incubated some of America's earliest open data portal ecosystems, unleashing private innovation by liberating hundreds of municipal datasets around domains like food safety violations, nurse visits, and building code complaints for reuse in grassroots apps (Mergel et al., 2018).

This preference for open participatory architecture advancing collaborative local priorities emanates from America's constitutional design fostering individual liberties and distributed checks on institutional power (see Bastos et al., 2022). Where Beijing perceives centralized control through technology and bureaucratic efficiency as requisites of national stability (US-China Economic and Security Review Commission, 2020), Washington treats pervasive surveillance and unchecked state leverage with skepticism rooted in founding beliefs around self-determination and democratic accountability (Wiig, 2016). Consequently, its smart city model favors bolstering horizontal connectivity across civic units through tools residents actively co-develop like distributed solar microgrids and community health monitoring apps rather than proprietary municipal automation by government partners (Shadowen et al., 2020).

America's smart city emphasis on participatory design thinking and open innovation also reconciles its private sector–led technology development model where firms like Microsoft, Cisco, and IBM dominate sophisticated enterprise solutions markets (Voorwinden, 2021). With public sectors averse toward large-scale internal automation that risks displacing jobs, bottom-up consumer platforms enabling peer exchange and entrepreneurship sustain tech penetration. Grassroots smart city emergence centered on local priorities thus aligns national economic architecture by empowering businesses shipping polished apps and analytics offerings competitive globally rather than enforcing state modernization drives (Reed & Keech, 2019). This organic market-based approach underlies the US model (see more in Perätalo et al., 2023; R. Yang & Zhen, 2024).

Chinese and American smart city convergences

Despite philosophical divergences rooted in political economies and governance histories, Sino-American smart city initiatives reveal areas of alignment, particularly in addressing common urban challenges (Hu & Zheng, 2021). As common pain points afflicting modern metropolises like congestion, soaring housing costs, and aged infrastructure obligations strain leadership worldwide, pragmatic considerations driving parallel policy responses hold the potential to bridge ideological divisions baked into Chinese and American smart urbanism models at municipal street levels even if national rivalry festers (also see Joss et al., 2019). Technology's embeddedness infiltrating all aspects of 21st-century urban life furnishes natural pressure for adversaries to cooperate ensuring common foundations function (e.g., Ro et al., 2023).

One zone of convergence comes with sustainability upgrades like renewable microgrids, electric vehicle fleets, and green buildings prioritized by every city plagued by air pollution and climate vulnerability (Marsal-Llacuna, 2020). As global consensus solidifies around essential decarbonization and resilience mandates, standards around issues like distributed solar storage, sensor-enabled humidity controls, and electric bus fleets achieve sufficient commercial maturity and political urgency for harmonious adoption even between rivals (Obringer & Nateghi, 2021). Global municipal climate networks allow Chinese and American cities to undertake joint renewable transitions assisted by shared technologies even while competing domestically (Dai et al., 2023). For instance, the C40 Cities Climate Leadership Group includes both Chinese cities like Beijing and Shenzhen and US cities like New York and Los Angeles. Through this network, these cities share best practices and technologies for renewable energy implementation (Lee & Meene, 2012). Another example is the Global Covenant of Mayors for Climate & Energy, which includes over 10,000 cities worldwide, including those from China and the US. This network facilitates knowledge exchange on sustainable energy solutions (Sporchia et al., 2023). Additionally, the US-China Climate-Smart/Low-Carbon Cities Summit, held in 2015 and 2016, brought together cities from both countries to share experiences and technologies in areas such as clean energy and sustainable transportation (*U.S.-China Climate-Smart Cities Initiative*, n.d.).

Another sphere primed for collaboration sits with epidemic response given existential public health risks from viruses ignoring borders (Pagani & Fracastoro, 2021). Early COVID-19 battles spotlighted domains like infection tracing platforms, vaccine coordination systems, and emergency messaging channels (Bayani et al., 2023; Mena Lora et al., 2023; Ricci et al., 2021) where joint Sino-American city development cooperation could furnish mutually beneficial tools no side could easily develop alone that save lives globally (Allam & Jones, 2020). Future pandemic threats may compel such joint preparedness despite countervailing national security logics (Söderström, 2021).

Chinese smart cities as international standards vehicles

While zones of modest convergence exist between Chinese and American smart cities bound by common municipal imperatives, Beijing also clearly perceives constructing sophisticated urban data and AI architectures abroad seeded by its own developers and protocols as a pivotal avenue for expanding international influence and shaping global technological standards to its advantage (Hoffman, 2022; Taylor, 2022, especially pp. 111–130; US-China Economic and Security Review Commission, 2020). Its Digital Silk Road program subsidizing exports of advanced platforms like "City Brain" complexes, safe city surveillance monitors, and sensor-laden new satellite cities expressly seeks to anchor Chinese supremacy over key building blocks essential for future infrastructure modernity worldwide (Caprotti & Liu, 2022; Kurlantzick & West, 2020).

Already, hundreds of Chinese-assembled video surveillance networks, monitored by AI-enhanced image processing suites, oversee streets from Islamabad to Johannesburg (Mazzocco, 2022; Sahin, 2020). These visible manifestations of technological systems create dependent path dynamics that likely privilege future compatibility upgrades and extensions from the original providers (Sahin, 2020). As cities worldwide may adopt either Chinese or American smart city technologies, including sensor networks, integrated command centers, cloud analytics, and AI-based optimization tools for core utility management, the differing technological ecosystems may present challenges for future integration and interoperability (Birman, 2024; K. Johnson, 2024).

In parallel, Beijing is actively working to shape international standards for next-generation telecommunications infrastructure as dependency on its 5G equipment manufacturers like Huawei deepens globally (Kim et al., 2020). For example, China submitted 830 technical documents related to wired communications specifications to the International Telecommunication

Union (ITU) in 2019, the most of any country and more than the next three – South Korea, the US, and Japan – combined (Ryugen & Akiyama, 2020). In the International Organization for Standardization (ISO), China expanded its participation in technical committees and subcommittees from 465 in 2005 to 668 in 2021, ranking third globally (Sheehan & Feldgoise, 2023). Furthermore, Chinese companies have become leading contributors to 5G standards (Becker et al., 2024; Rühlig, 2023). Huawei is the top filer of standard-essential patents for 5G, according to the Tokyo-based Cyber Creative Institute, and leads in 5G-related contributions to 3GPP (3rd Generation Partnership Project), an international organization that develops telecom standards, surpassing European rivals and US-based Qualcomm (Ryugen & Akiyama, 2020). Collectively, Chinese companies hold about one-third of the world's 5G-related standard-essential patents (H. Lee, 2021).

The technical architecture of ultrafast metropolitan connectivity, prizing seamless sensor integration and millisecond responsiveness, carries an embedded geopolitics footprint as Chinese firms expand their market share (Zúñiga et al., 2024). This is particularly evident in emerging technological sectors connected through 5G networks, such as autonomous cars, AI, smart factories, and smart cities (H. Lee, 2021). Such dynamics spotlight Beijing's acute recognition of smart cities' significant role in anchoring its preferences within contested technology domains through first-mover advantage and potential stickiness effects (Kim et al., 2020). Moreover, its standardization efforts are part of a broader strategy, reportedly including a medium-term plan nicknamed "China Standards 2035", which complements the "Made in China 2025" industrial modernization plan (Ryugen & Akiyama, 2020). This strategy aims not just at dominance in production but also at controlling the rules governing emerging technologies (H. Lee, 2021). However, it is also important to note that while China's influence in standard-setting is growing, the international standards system remains largely robust, with decisions in reputable standards development organizations typically being consensus-driven and non-binding (Sheehan & Feldgoise, 2023).

US cities exporting sociotechnical governance models

Meanwhile, the US also views its own smart city innovation epicenters like New York, Los Angeles, and Chicago as influential demonstration cases for responsible technology integration, advancing American interests abroad (Zúñiga et al., 2024, p. 1644). These cities are developing digitally enhanced, data-rich civic management pilots that aim to balance innovation with citizen engagement and privacy concerns (Bustamante & Calidas, 2024; Q. Hu & Zheng, 2021). By creating urban environments that integrate public participation with accountability into complex automation and predictive analytics systems, American models promote sociotechnical compromises that balance technological advancement with democratic ideals (Walters, 2011). This approach reflects the US's traditional emphasis on market-driven solutions and minimal government intervention in standard-setting (Zúñiga et al., 2024, p. 1639).

Cities across other developed countries are observing and sometimes emulating governance innovations emerging from leading US urban test beds. These innovations often involve reconciling competing priorities around controversial technologies like facial recognition, predictive policing algorithms, and AI systems in public sectors. In many cases, concerned residents and experts collaborate with city administrators to continuously refine oversight regimes for advanced applications, creating templates that balance community needs with technological progress (Zúñiga et al., 2024, p. 1640). Major US technology providers like Microsoft, IBM, and Cisco market their vision for smart cities globally, emphasizing flexible, open, and integrative cloud-hosted solutions (Zúñiga et al., 2024, p. 1640). By promoting modular approaches to refashioning municipal operations, these American firms present smart cities as collaborative environments secured by design rather than through pervasive surveillance. This approach contrasts with more centralized models promoted by some other countries (Q. Hu & Zheng, 2021).

Yet, it is important to note that while the US has historically led in setting global technology standards, it is now facing increased competition, particularly from China. The US government is recognizing the need for a more proactive approach to standard-setting to maintain its influence in this arena (Zúñiga et al., 2024, pp. 1641–1642).

Geopolitical implications and global adoption of smart city models

The competition between Chinese and American approaches to smart cities extends far beyond urban development, representing a crucial battleground in the broader geopolitical contest for technological supremacy and global influence (Ekman, 2019). This rivalry reflects divergent visions for the future international order, with smart city initiatives serving as strategic instruments of foreign policy for both nations (Q. Hu & Zheng, 2021; Joss et al., 2019). The global race to develop and export smart city models has significant implications for international relations, economic development, and the balance of power in the digital age (Datta & Odendaal, 2019).

China's ambitious Digital Silk Road and Belt and Road Initiative leverage smart city technologies to expand Beijing's influence across Eurasia and beyond (Hillman, 2021). This strategy not only aims to cement pan-Eurasian infrastructural and institutional integration through Chinese technologies but also seeks to influence international legal norms and standards governing these technologies (Hussain et al., 2024; C. Zhang et al., 2022). As cities adopt Chinese technologies like Beidou satellites for geospatial analytics or

Alibaba Cloud for civic data processing, they may also be adopting Chinese-influenced legal and regulatory frameworks, potentially shifting the global legal landscape toward a more Sino-centric order.

The US, recognizing the strategic importance of maintaining its technological and legal leadership, has responded with initiatives aimed at promoting its own vision of smart cities. These efforts seek to counter Chinese influence by showcasing American approaches to urban innovation that emphasize individual privacy, pluralistic oversight, and protection of civil liberties – values that are deeply embedded in the US legal tradition (Q. Hu & Zheng, 2021). The US strategy aims to preserve its post–Cold War ideological and legal influence, which could be undermined if authoritarian governance models become embedded in the urban fabric and legal systems of developing nations through Chinese-led smart city projects (Zúñiga et al., 2024).

In light of this, the global adoption of smart city models has significant implications for international law. As urban areas become increasingly networked and data-driven, the ability to shape the legal and regulatory frameworks for these systems becomes a critical source of influence. This has led to intensified competition in international standard-setting bodies and legal forums, with both China and the US seeking to ensure their preferred approaches become global norms. The outcome of this contest will have far-reaching implications for international law, particularly in areas such as data protection, privacy rights, intellectual property, and the balance between state sovereignty and individual rights in digital spaces. Moreover, the development of smart cities is challenging existing international legal frameworks. Issues such as cross-border data flows, cybersecurity, and the application of traditional legal concepts to new technological realities are forcing a reconsideration of established international legal norms. The way these legal challenges are resolved will have profound implications for the future of global governance in the digital age.

The role of cities in shaping the norms and rules of smart city governance

While the previous section attended to the geopolitical competition between China and the US in smart city development, this section shifts attention to the role of cities themselves in shaping the norms and rules of smart city governance. As national governments deal with rapid technological changes, cities are emerging as key players in setting standards, asserting governance autonomy, and navigating the challenges of regulating global tech platforms. This section explores how cities are leveraging their growing influence to shape international law and policy in the realm of smart city development, offering a counterpoint to the state-centric narratives of geopolitical competition.

Setting global standards

As national governments struggle to respond to the fast technological change redefining urban infrastructure and service delivery models, the world's great cities are banding together into formal standard-setting alliances and knowledge-sharing networks aiming to codify common governance priorities (Araya, 2015). Groups like the C40 Cities, Smart Cities Network, and United for Smart Sustainable Cities consolidate influential municipal leaders to articulate consensus positions on issues like data ethics, citizen participation, and environmental sustainability that later feed into programs at institutions from the United Nations to WTO shaping critical rules of international exchange (Estevez et al., 2021; Lai et al., 2020; Naiki, 2024). A prime example of this trend is the Cities' Coalition for Digital Rights (CCDR), an international network of cities formed in 2018 by the city councils of Barcelona, Amsterdam, and New York. The CCDR, which has grown to encompass 60 cities, aims to promote citizens' digital rights globally (Cities Coalition for Digital Rights, n.d.). This coalition aligns with UN-Habitat's flagship program called "People-Centered Smart Cities", which advocates for inclusiveness while subverting the technocratic smart city paradigm (Calzada et al., 2023). By focusing on "the right to have digital rights", the CCDR and its member cities are establishing new norms and expectations for how technology should be integrated into urban life, providing a model that could inform broader international agreements and standards related to smart cities and digital governance (Calzada et al., 2023).

In effect, municipal coalitions crucially furnish channels for more fragmented cities, overwhelmed by the pace of innovation shifts, to harmonize around shared policy and technical protocol challenges related to smart service integration (also see Calzada et al., 2023). By collectively detailing concerns like safe mobility systems requirements for autonomous vehicles, transparency principles for AI-enhanced administrative algorithms, and interoperability specifications for real-time utility data integration, urban blocs can more decisively advance their interests within complex multilateral negotiations and public-private regulatory deliberative bodies struggling to govern rapid technology change (Collier, 2021; Mookerjee, 2021). For example, the National Smart Coalitions Partnership unites regional efforts to advance smart technology solutions across cities in the US, encouraging collaboration and sharing best practices to lower barriers to deploying, replicating, and scaling smart solutions (National Smart Coalitions Partnership, n.d.). Similarly, the Smart Surfaces Coalition demonstrates the value of smart surfaces that manage sun and rain, helping cities reduce extreme heat, minimize flooding, and improve public health, thereby creating more livable urban environments (Smart Surfaces Coalition, n.d.). These coalitions enable cities to collectively address shared challenges and develop standardized approaches to emerging technologies, enhancing their ability to influence international standards and regulatory frameworks (Smart Surfaces Coalition, n.d.). By participating in these networks, cities can better navigate the complexities of technological change and ensure that their governance priorities are reflected in global policy discussions (also see Mills et al., 2021).

In addition, urban technology dialogues facilitated by cities like New York and Singapore have allowed for detailed discussions on security properties for IoT sensor grids and other smart city technologies. These dialogues involve municipal technology departments, platform vendors, and civil society experts, contributing to the development of international standards. For instance, the U.S.-Singapore Critical and Emerging Technology Dialogue has committed to deepening information-sharing and consultations on international AI security, safety, trust, and standards development (The White House, 2023). This collaboration includes the launch of a bilateral AI Governance Group focused on advancing shared principles for safe, trustworthy, and responsible AI innovation (The White House, 2023).

In technology domains like autonomous vehicles, urban coalitions are also forming to address governance challenges (Aoyama & Leon, 2021). New York City's participation in the Connected Vehicle Pilot Deployment Program demonstrates how cities can lead in implementing and shaping standards for emerging urban technologies (e.g., US Department of Transportation. ITS Joint Program Office & Transportation, 2017). As cities like Amsterdam, Helsinki, New York, and Singapore take the lead in urban AI initiatives, they are increasingly influencing global governance processes. Their recommendations often feed into programs at major international institutions, potentially shaping critical rules of international exchange (Townsend et al., 2023). As smart services become more integrated into urban environments, coordinated city leadership voices are becoming essential in establishing guidelines and promoting responsible innovation through formal international standard channels (Townsend et al., 2023).

Asserting governance autonomy

In parallel to coordinating standards priorities through networks, individual major cities are also more unilaterally leveraging growing economic and political clout to take direct governance control over digitization initiatives launched within their borders that impact residents and businesses (Foth et al., 2022). Through local ordinances, procurement stipulations, and pilot operating agreements, urban administrations are moving toward data sovereignty frameworks retaining and regulating access mediation rights over rich information streams like mobility patterns, power loads, and environmental indicators originating from civic environments (Cuno et al., 2019; J. Johnson et al., 2022).

For example, Barcelona has enacted a comprehensive data governance strategy rooted in principles of transparency, ethical AI use, and accountability. This strategy is enforced through new municipal initiatives, including the creation of a public register of algorithms used by the city administration and the implementation of audits to assess the risks of these algorithms (Barcelona Digital City, 2021). The city has established the governing principles for technological applications, encompassing human action and supervision, technical robustness and security, data privacy and governance, transparency and information, diversity, inclusion and fairness, social and environmental commitment, and accountability and democratic control (Barcelona Digital City, 2021). Barcelona has also developed a set of instructions establishing protocols for AI-based systems' design and implementation, created a technical guide detailing processes for responsible AI use, and included digital rights clauses in the tendering of AI-based solutions (Cerrillo-Martínez & Casadesús-de-Mingo, 2021).

By directly mediating data-sharing practices and intelligent applications from the outset of procurement, proactive authorities like Barcelona are establishing new standards for safeguarding resident interests. This approach aims to balance the need for innovation with the protection of citizens' rights and data sovereignty, addressing concerns that may be overlooked by national regimes more focused on rapid technological advancement and economic growth (Lynch, 2020). These efforts align with broader initiatives such as the CCDR, which Barcelona cofounded, promoting digital rights in urban contexts globally (Barcelona Digital City, 2021). Such approaches demonstrate how cities are taking a leading role in shaping the governance of urban data spaces, emphasizing the importance of data sovereignty and ethical use of technology in smart city development (Cerrillo-Martínez & Casadesús-de-Mingo, 2021).

Regulating tech platforms

In addition to asserting smart governance autonomy, cities are now some of the first regulators dealing with global tech platforms for providing essential services and infrastructure (Ranchordás & Goanta, 2020). This role puts cities in a unique position to influence international norms and practices regarding data privacy, algorithmic transparency, and digital rights. Indeed, it is emblematic of a broader trend in the governing landscape. That is, cities have become an increasingly powerful regulatory player and are more inclined to exert control over digital environments that primarily influence people's everyday lives (Vanolo, 2013). One prime case in point is New York City's approach to addressing automated decision systems (ADS). In 2018, the city passed Local Law 49, establishing a task force to examine the use of ADS in city government and make recommendations for their regulation

(Automated Decision Systems Task Force, 2019). To a large extent, this initiative has set a precedent for other cities worldwide, demonstrating how local governance can shape the implementation of AI and machine learning technologies in urban environments (Automated Decision Systems Task Force, 2019; also see Baykurt, 2022). The task force's work has highlighted the complex interplay between technological innovation, public administration, and civil rights, addressing issues such as bias, transparency, and accountability in algorithmic decision-making (Baykurt, 2022).

Besides, the influence of these local initiatives stretches far beyond city limits. As Calzada (Calzada, 2018a) argues, city-level regulation of digital technologies can serve as a laboratory for democratic experimentation, potentially influencing national and international regulatory frameworks. This is particularly evident in the case of facial recognition technology, where San Francisco's 2019 ban on its use by city agencies sparked a nationwide debate and influenced similar bans in other U.S. cities (Conger et al., 2019). These local actions collectively shape the global landscape of tech regulation, potentially influencing international legal frameworks and corporate practices worldwide (Hernández-Morales, 2023; Qandeel, 2024). Further, cities are using their purchasing power to pressure tech companies in a way that is growing. For instance, Amsterdam (City of Amsterdam Algorithm Register, n.d.) and Helsinki (City of Helsinki AI Register, n.d.) have launched AI Registries, providing overviews of the AI systems and algorithms used by their respective city governments (also see K. Johnson, 2020). This initiative promotes transparency and accountability in the use of AI technologies, setting a standard that could influence international norms on algorithmic governance.

Inventing legal and regulatory approaches

Last but not least, smart cities serve as real-world laboratories for testing new legal and regulatory approaches to emerging technologies (Decker, 2016). This experimentation allows cities to develop innovative governance models that can inform national and international policy-making. Recent years have seen the growing recognition of the notion of cities as "urban laboratories", with scholars like Karvonen and Heur (2014) arguing that cities provide unique spaces for experimentation with new forms of governance and technological innovation.

The case of Toronto's Quayside project, despite its eventual cancellation, provides a compelling example of this phenomenon. The project, a partnership between Waterfront Toronto and Sidewalk Labs (a subsidiary of Alphabet Inc.), aimed to create a technologically advanced neighborhood that would serve as a model for future urban development. Central to the project was a proposed data trust model, which aimed to ensure that urban data would be used for the public good (Chantry, 2023; Nelischer, 2024). Although the project was ultimately abandoned, the debates and proposed solutions it generated continue to influence smart city governance discussions worldwide (Haggart & Spicer, 2022). The Quayside project highlighted the complex challenges of data governance in smart city contexts. These issues are at the heart of ongoing debates about the legal and ethical implications of smart city technologies, and Toronto's experience has provided valuable insights for other cities facing similar challenges (Spicer & Zwick, 2021).

Singapore's Smart Nation initiative offers another example of a city serving as a laboratory for legal innovation in the smart city context, being at the forefront of developing a comprehensive legal and regulatory framework for smart city technologies (Frana, 2024). In particular, Singapore's Personal Data Protection Act (PDPA) and its complementary Model AI Governance Framework provide models for other cities and countries grappling with similar issues (Ministry of Digital Development and Information, n.d.). The PDPA, enacted in 2012 and subsequently updated, establishes a baseline standard for data protection in Singapore, while the AI Governance Framework offers guidelines for the responsible development and deployment of AI systems. A key feature of Singapore's approach is the creation of "regulatory sandboxes" that relax regulatory requirements for innovative companies (Frana, 2024). These sandboxes allow companies to test emerging AI solutions and speed them to market while remaining under the watchful eye of government agencies (Infocomm Media Development Authority, 2023a; Laurent et al., 2021). This approach demonstrates how cities can lead in developing flexible legal frameworks that balance innovation with privacy and security concerns. Singapore has also taken steps to address the emerging challenges of generative AI. In 2023, the Infocomm Media Development Authority launched a first-of-its-kind generative AI evaluation sandbox in collaboration with the AI Verify Foundation (Infocomm Media Development Authority, 2023b). This initiative aims to provide a trusted environment for testing and evaluating generative AI systems, addressing concerns about their reliability and potential risks.

Concluding remarks

This chapter has explored the intersection of smart cities and the geopolitics of international law, examining why and how smart urbanization has emerged as a major new arena of techno-strategic competition between states. Several key arguments and findings have crystallized through analysis.

First, the rapid development of sensor-laden urban environments, data-driven civic management systems, and AI-powered municipal automation infrastructure over the past decade has opened up a decisive new playing field where major powers like China and the US clash over technological leadership and models of technology governance. Fusing cities with next-generation information and communication technology ecosystems confers advantages spanning economic dynamism, industrial upgrading, scientific innovation, commercial dominance, and military capability that are essential for comprehensive national power and influence in the digital age.

Consequently, states are actively employing smart city-building within explicitly geostrategic foreign policy frameworks aimed at expanding national strengths while promoting their own governance values and standards internationally. As extensively covered through the China-US case study, both countries are leveraging state-backed investments under different initiatives to entrench global supply chains, infrastructure connectivity, and regulatory architectures aligning with their interests. They recognize smart cities' geopolitical instrumentality.

In addition, the chapter spotlights cities themselves as increasingly pivotal voices shaping smart development governance. Through global networks and coalitions like the CCDR, cities are asserting their autonomy in governing digital technologies and setting standards that influence international policy. By pooling best practices around shared priorities like sustainability and digital rights, and forging consensus positions to feed into formal multilateral processes, city blocs are shaping common rules for managing transformative domains undersupplied by existing legal institutions. Initiatives such as Barcelona's comprehensive data governance strategy, New York City's task force on ADS, and Singapore's regulatory sandboxes for AI demonstrate how cities are taking proactive steps to regulate emerging technologies and protect citizens' rights. These local actions collectively contribute to shaping the global landscape of tech regulation, potentially influencing international legal frameworks and corporate practices worldwide.

These findings carry significant implications for global smart city development and associated international legal orders. They necessitate greater recognition of cities as transnational governance stakeholders in their own right rather than just passive sites of strategic competition between states and companies. Their amplifying voices require accommodation within critical regulatory conversations shaping digital transitions inside formal institutions that still marginalize urban participation. Cities themselves must also pursue greater consolidation around issues of common concern like responsible technology procurement and data stewardship through transmunicipal pacts if their preoccupations are to decisively imprint the global scale.

Future research should focus extensively on elaborating on exact mechanisms facilitating enhanced city-to-city collaboration and tangibly inserting heightened urban perspectives into pivotal debates like AI standard setting and platform accountability through legal venues cities can meaningfully access. More importantly, further research is needed to explore how the evolving landscape of smart cities will continue to shape and be shaped by international law. Key areas for future investigation include the long-term impacts of divergent smart city models on global legal norms, the effectiveness of city-led initiatives in influencing international standards, and the potential for new forms of global governance that can balance innovation, security, and individual rights in the smart cities of the future.

References

Acuto, M. (2013). *Global cities, governance and diplomacy: The urban link.* Routledge.

Al-Huthaifi, R., Li, T., Huang, W., Gu, J., & Li, C. (2023). Federated learning in smart cities: Privacy and security survey. *Information Sciences*, *632*, 833–857. https://doi.org/10.1016/j.ins.2023.03.033

Albino, V., Berardi, U., & Dangelico, R. M. (2015). Smart cities: Definitions, dimensions, performance, and initiatives. *Journal of Urban Technology*, *22*(1), 3–21. https://doi.org/10.1080/10630732.2014.942092

Allam, Z., & Jones, D. S. (2020). On the coronavirus (COVID-19) outbreak and the smart city network: Universal data sharing standards coupled with artificial intelligence (AI) to benefit urban health monitoring and management. *Healthcare*, *8*(1), 1–9. https://doi.org/10.3390/healthcare8010046

Amsellem, A. (2021). The noise of silent machines: A case study of LinkNYC. *Smart Surveillance*, *19*(2), 168–186. https://doi.org/10.24908/ss.v19i2.14302

Anderson, C., & Jung, J. K. (2023). For a cooperative "smart" city yet to come: Place-based knowledge, commons, and prospects for inclusive municipal processes from Seattle, Washington. *Urban Planning*, *8*(2), 6–16. https://doi.org/10.17645/up.v8i2.6597

Anghie, A. (2005). *Imperialism, sovereignty and the making of international law.* Cambridge University Press.

Aoyama, Y., & Leon, L. F. A. (2021). Urban governance and autonomous vehicles. *Cities*, *119*(December), 103410. https://doi.org/10.1016/j.cities.2021.103410

Araya, D. (2015). Smart cities and the network society: Toward commons-driven governance. In D. Araya (Ed.), *Smart cities as democratic ecologies* (pp. 11–22). Palgrave Macmillan. https://doi.org/10.1057/9781137377203_2

Aust, H.P., & Nijman, J. E. (2023). *The urbanisation of international law – A post-cold war history* (No. 2023-31; Amsterdam Law School Research Paper). https://ssrn.com/abstract=4584309

Aust, H.P., & Nijman, J. E. (Eds.). (2021). *Research handbook on international law and cities.* Edward Elgar Publishing.

Automated Decision Systems Task Force. (2019, November). *New York City automated decision systems task force report.* https://www.nyc.gov/assets/adstaskforce/downloads/pdf/ADS-Report-11192019.pdf

Badran, A. (2023). Developing smart cities: Regulatory and policy implications for the state of Qatar. *International Journal of Public Administration*, *46*(7), 519–532. https://doi.org/10.1080/01900692.2021.2003811

Barcelona Digital City. (2021). *The municipal algorithms and data strategy for an ethical promotion of artificial intelligence.* https://ajuntament.barcelona.cat/digital/en/technology-accessible-everyone/ethical-use-artificial/ethical-use-artificial-intelligence/municipal

Barns, S. (2020). *Platform urbanism: Negotiating platform ecosystems in connected cities*. Palgrave Macmillan. 10.1007/978-981-32-9725-8

Bastos, D., Fernández-Caballero, A., Pereira, A., & Rocha, N. P. (2022). Smart city applications to promote citizen participation in city management and governance: A systematic review. *Informatics*, *9*(4), 1–29. https://doi.org/10.3390/informatics 9040089

Bateman, J. (2022). *U.S.-China technological "Decoupling": A strategy and policy framework*. Carnegie Endowment for International Peace. https://carnegieen dowment.org/files/Bateman_US-China_Decoupling_final.pdf

Bayani, F., Hashkavaei, N. S., Arjmand, S., Rezaei, S., Uskoković, V., Alijanianzadeh, M., Uversky, V. N., Ranaei Siadat, S. O., Mozaffari-Jovin, S., & Sefidbakht, Y. (2023). An overview of the vaccine platforms to combat COVID-19 with a focus on the subunit vaccines. *Progress in Biophysics and Molecular Biology*, *178* (January), 32–49. https://doi.org/10.1016/j.pbiomolbio.2023.02.004

Baykurt, B. (2022). Algorithmic accountability in U.S. cities: Transparency, impact, and political economy. *Big Data & Society*, *9*(2), 20539517221115424. https:// doi.org/10.1177/20539517221115426

Becker, C., Oever, N. ten, & Nanni, R. (2024). Interrogating the standardisation of surveillance in 5G amid US–China competition. *Information, Communication & Society*, *27*(16), 2834–2851. https://doi.org/10.1080/1369118X.2024.2302991

Birman, S. (2024, July 10). *The intensifying U.S. – China tech race: Grasping for power in the twenty-first century*. The Science Survey. https://thesciencesurvey. com/news/2024/07/10/the-intensifying-u-s-china-tech-race-grasping-for-power-in-the-twenty-first-century/

Brown, T. E. (2019). Human rights in the smart city: Regulating emerging technologies in city places. In L. Reins (Ed.), *Regulating new technologies in uncertain times* (pp. 47–65). T.M.C. Asser Press. https://doi.org/10.1007/978-94-6265-279-8_4

Bustamante, C. M. V., & Calidas, D. V. (2024, February). *Unraveling the political dynamics shaping the U.S. strategy for technology leadership*. Belfer Center for Science and International Affairs Working Paper. https://www.belfercenter.org/ sites/default/files/files/publication/HKS_USStrategyTechLeadership_v4.pdf

Calzada, I. (2018a). (Smart) citizens from data providers to decision-makers? The case study of Barcelona. *Sustainability*, *10*(9), 3252. https://doi.org/10.3390/ su10093252

Calzada, I. (2018b). From smart cities to experimental cities? In V. M. B. Giorgino & Z. Walsh (Eds.), *Co-designing economies in transition* (pp. 191–217). Palgrave Macmillan. https://doi.org/10.1007/978-3-319-66592-4_11

Calzada, I., Pérez-Batlle, M., & Batlle-Montserrat, J. (2023). People-centered smart cities: An exploratory action research on the Cities' Coalition for Digital Rights. *Journal of Urban Affairs*, *45*(9), 1537–1562. https://doi.org/10.1080/07352166. 2021.1994861

Caprotti, F., & Liu, D. (2022). Platform urbanism and the Chinese smart city: The co-production and territorialisation of Hangzhou City Brain. *GeoJournal*, *87*(3), 1559–1573. https://doi.org/10.1007/s10708-020-10320-2

Cerrillo-Martínez, A., & Casadesús-de-Mingo, A. (2021). Data governance for public transparency. *Profesional de La Información*, *30*(4), e300402. https://doi. org/10.3145/epi.2021.jul.02

Chantry, W. (2023). 'Built from the internet up': Assessing citizen participation in smart city planning through the case study of Quayside, Toronto. *GeoJournal*, *88*(2), 1619–1637. https://doi.org/10.1007/s10708-022-10688-3
Chimni, B. S. (2018). Customary international law: A third world perspective. *American Journal of International Law*, *112*(1), 1–46. https://doi.org/10.1017/ajil.2018.12
Choenni, S., Bargh, M. S., Busker, T., & Netten, N. (2022). Data governance in smart cities: Challenges and solution directions. *Journal of Smart Cities and Society*, *1*(1), 31–51.
Cities Coalition for Digital Rights. (n.d.). *About us*. Retrieved July 20, 2024, from https://citiesfordigitalrights.org/thecoalition
City of Amsterdam Algorithm Register. (n.d.). *What is the algorithm register?* Retrieved July 20, 2024, from https://algoritmeregister.amsterdam.nl/en/ai-register/
City of Helsinki AI Register. (n.d.). *What is AI register?* Retrieved July 20, 2024, from https://ai.hel.fi/en/ai-register/
Collier, C. (2021, October 8). *Smart regions unite to form the National Smart Coalitions Partnership*. Smart Cities Connect. https://smartcitiesconnect.org/smart-regions-unite-to-form-the-national-smart-coalitions-partnership/
Conger, K., Fausset, R., & Kovaleski, S. F. (2019, May 14). *San Francisco bans facial recognition technology*. The New York Times. https://www.nytimes.com/2019/05/14/us/facial-recognition-ban-san-francisco.html
Correia, D., Marques, J. L., & Teixeira, L. (2022). The state-of-the-art of smart cities in the European Union. *Smart Cities*, *5*(4), 1776–1810. https://doi.org/10.3390/smartcities5040089
Criekemans, D. (2010). Regional sub-state diplomacy from a comparative perspective: Quebec, Scotland, Bavaria, Catalonia, Wallonia and Flanders. *The Hague Journal of Diplomacy*, *5*(1–2), 37–64. https://doi.org/10.1163/1871191x-05010103
Crouch, C., & Galès, P. Le. (2012). Cities as national champions? *Journal of European Public Policy*, *19*(3), 405–419. https://doi.org/10.1080/13501763.2011.640795
Cugurullo, F. (2020). Urban artificial intelligence: From automation to autonomy in the smart city. *Frontiers in Sustainable Cities*, *2*(July), 1–14. https://doi.org/10.3389/frsc.2020.00038
Cuno, S., Bruns, L., Tcholtchev, N., Lämmel, P., & Schieferdecker, I. (2019). Data governance and sovereignty in urban data spaces based on standardized ICT reference architectures. *Data*, *4*(1), 1–24. https://doi.org/10.3390/data4010016
Curtis, S. (2016). *Global cities and global order*. Oxford University Press. https://doi.org/10.1017/CBO9781107415324.004
Dai, F., Bedsworth, L., Lewis, J., Gordon, J., & Edwards, L. (2023). *Subnational climate action in the U.S. and China: Where we are and opportunities for cooperation*. https://ccci.berkeley.edu/sites/default/files/10_Subnational%2520Climate%2520Action%2520In%2520the%2520United%2520States%2520in%2520the%2520U.S%2520and%2520China%2520Where%2520We%2520Are%2520and%2520Opportunities%2520for%2520Cooperation_English.pdf
Datta, A., & Odendaal, N. (2019). Smart cities and the banality of power. *Environment and Planning D: Society and Space*, *37*(3), 387–392. https://doi.org/10.1177/0263775819841765

Decker, A. (2016). Smart law for smart cities. *Fordham Urban Law Journal*, *41*(5), 1491–1498.

Dezalay, Y., & Garth, B. G. (2002). *The internationalization of palace wars: Lawyers, economists, and the contest to transform Latin American states*. The University of Chicago Press.

Ekman, A. (2019, December). China's smart cities: The new geopolitical battleground. *French Institute of International Relations*. https://www.ifri.org/sites/default/files/atoms/files/ekman_smart_cities_battleground.pdf

Eldem, T. (2021, July 20). Canal Istanbul: Turkey's controversial megaproject. *SWP Comment*. https://www.swp-berlin.org/10.18449/2021C43/

Estevez, E., Cenci, K., Fillottrani, P., & Janowski, T. (2021). Review of international standards and policy guidelines for smart sustainable cities. In E. Estevez, T. A. Pardo, & H. J. Scholl (Eds.), *Smart cities and smart governance* (pp. 69–99). Springer. https://doi.org/10.1007/978-3-030-61033-3_4

Faraji, S. J., Zanganeh, A., Pirastefar, P., & Hedari, H. (2023). Exploring the effects of key factors of smart urban governance on geocultural relations. *Geopolitics Quarterly*, *19*(2), 345–369. https://doi.org/20.1001.1.17354331.1402.19.70.14.7

Foth, M., Anastasiu, I., Mann, M., & Mitchell, P. (2022). From automation to autonomy: Technological sovereignty for better data care in smart cities. In B. T. Wang & C. M. Wang (Eds.), *Automating cities: Design, construction, operation and future impact* (pp. 319–343). Springer. https://doi.org/10.1007/978-981-15-8670-5_13

Frana, P. L. (2024). Assessing Smart Nation Singapore as an international model for AI responsibility. *International Journal on Responsibility*, *7*(1), 1–31. https://doi.org/10.62365/2576-0955.1109

Getachew, A. (2019). *Worldmaking after Empire: The rise and fall of self-determination*. Princeton University Press.

Gilman, H. R. (2016). *Participatory budgeting and civic tech: The revival of citizen engagement*. Georgetown University Press.

GlobalData. (2022, November 23). *Smart cities market size, share, trends, analysis and forecast by region, application, infrastructure, grid and segment forecast to 2026*. https://www.globaldata.com/store/report/smart-cities-market-analysis/

Große-Bley, J., & Kostka, G. (2021). Big data dreams and reality in Shenzhen: An investigation of smart city implementation in China. *Big Data & Society*, *8*(2). https://doi.org/10.1177/20539517211045171

Haggart, B., & Spicer, Z. (2022). Infrastructure, smart cities and the knowledge economy: Lessons for policymakers from the Toronto Quayside project. *Canadian Public Administration*, *65*(2), 295–313. https://doi.org/10.1111/capa.12460

Harvey, D. (2005). *A brief history of neoliberalism*. Oxford University Press.

He, W., Li, W., & Deng, P. (2022). Legal governance in the smart cities of China: Functions, problems, and solutions. *Sustainability*, *14*(15), 9738. https://doi.org/10.3390/su14159738

He, Y. (2024). Chinese digital platform companies' expansion in the Belt and Road countries. *The Information Society*, *40*(2), 96–119. https://doi.org/10.1080/01972243.2024.2317058

Heeks, R., Ospina, A. V., Foster, C., Gao, P., Han, X., Jepson, N., Schindler, S., & Zhou, Q. (2024). China's digital expansion in the Global South: Systematic literature review and future research agenda. *Information Society*, *40*(2), 69–95. https://doi.org/10.1080/01972243.2024.2315875

Henderson, C. (2021). Land grabs reexamined: Gulf Arab agro-commodity chains and spaces of extraction. *Environment and Planning A*, *53*(2), 261–279. https://doi.org/10.1177/0308518X20956657

Hernández-Morales, A. (2023). *Living cities: The battle over urban facial recognition*. POLITICO. https://www.politico.eu/newsletter/global-policy-lab/living-cities-the-battle-over-urban-facial-recognition/

Hillman, J. E. (2021). *The digital Silk Road: China's quest to wire the world and win the future*. Harper Business.

Hoffman, S. (2022). China's tech-enhanced authoritarianism. *Journal of Democracy*, *33*(2), 76–89. https://doi.org/10.1353/jod.2022.0019

Hu, Q., & Zheng, Y. (2021). Smart city initiatives: A comparative study of American and Chinese cities. *Journal of Urban Affairs*, *43*(4), 504–525. https://doi.org/10.1080/07352166.2019.1694413

Hu, R. (2019). The state of smart cities in China: The case of Shenzhen. *Energies*, *12*(22). https://doi.org/10.3390/en12224375

Hulle, I. Van. (2016). Grotius, informal empire and the conclusion of unequal treaties. *Grotiana*, *37*(1), 43–60. https://doi.org/10.1163/18760759-03700003

Hussain, F., Hussain, Z., Khan, M. I., & Imran, A. (2024). The digital rise and its economic implications for China through the Digital Silk Road under the Belt and Road Initiative. *Asian Journal of Comparative Politics*, *9*(2), 238–253. https://doi.org/10.1177/20578911231174731

Infocomm Media Development Authority. (2023a). *Data regulatory sandbox*. https://www.imda.gov.sg/how-we-can-help/data-innovation/data-regulatory-sandbox

Infocomm Media Development Authority. (2023b). *First of its kind generative AI evaluation sandbox for trusted AI by AI verify foundation and IMDA*. https://www.imda.gov.sg/resources/press-releases-factsheets-and-speeches/press-releases/2023/generative-ai-evaluation-sandbox

Ismagilova, E., Hughes, L., Dwivedi, Y. K., & Raman, K. R. (2019). Smart cities: Advances in research—An information systems perspective. *International Journal of Information Management*, *47*, 88–100. https://doi.org/10.1016/j.ijinfomgt.2019.01.004

Javed, A. R., Shahzad, F., Rehman, S., Zikria, Y. Bin, Razzak, I., Jalil, Z., & Xu, G. (2022). Future smart cities: Requirements, emerging technologies, applications, challenges, and future aspects. *Cities*, *129*, 103794. https://doi.org/10.1016/j.cities.2022.103794

Jiang, S. (2020). Marching to an international law powerhouse through cities: an ideational reflection on China's 'Belt and Road' Initiative. In G. Martinico & X. Wu (Eds.), *A Legal Analysis of the Belt and Road Initiative* (pp. 221–251). Palgrave Macmillan.

Johnson, J., Hevia, A., Yergin, R., Karbassi, S., & Levine, A. (2022). Data governance framework for smart cities: Key considerations for data management and use. *Journal of Law and Mobility*, 1–27.

Johnson, K. (2020, September 28). *Amsterdam and Helsinki launch algorithm registries to bring transparency to public deployments of AI*. VentureBeat. https://venturebeat.com/ai/amsterdam-and-helsinki-launch-algorithm-registries-to-bring-transparency-to-public-deployments-of-ai/

Johnson, K. (2024, July 4). *A global smart-city competition highlights China's rise in AI*. Wired. https://wired.me/technology/global-smart-city-competition-highlights-chinas-rise-in-ai/

Joss, S., Cook, M., & Dayot, Y. (2017). Smart cities: Towards a new citizenship regime ? A discourse analysis of the British smart city standard. *Journal of Urban Technology*, *24*(4), 29–49. https://doi.org/10.1080/10630732.2017.1336027

Joss, S., Sengers, F., Schraven, D., Caprotti, F., & Dayot, Y. (2019). The smart city as global discourse: Storylines and critical junctures across 27 cities. *Journal of Urban Technology*, *26*(1), 3–34. https://doi.org/10.1080/10630732.2018.1558387

Ju, J., Liu, L., & Feng, Y. (2024). Governance mechanism of public-private partnerships for promoting smart city performance: A multi-case study in China. *Cities*, *153*(October), 105295. https://doi.org/10.1016/j.cities.2024.105295

Kania, E. B. (2019, November 7). Securing our 5G future: The competitive challenge and considerations for U.S. policy. *Center for a New American Security*. https://www.cnas.org/publications/reports/securing-our-5g-future

Karvonen, A., & Heur, B. van. (2014). Urban laboratories: Experiments in reworking cities. *International Journal of Urban and Regional Research*, *38*(2), 379–392. https://doi.org/10.1111/1468-2427.12075

Kim, M., Lee, H., & Kwak, J. (2020). The changing patterns of China's international standardization in ICT under techno-nationalism: A reflection through 5G standardization. *International Journal of Information Management*, *54*(October), 102145. https://doi.org/10.1016/j.ijinfomgt.2020.102145

Kitchin, R., Coletta, C., Evans, L., & Heaphy, L. (2018). Creating smart cities. In C. Coletta, L. Evans, L. Heaphy, & R. Kitchin (Eds.), *Creating smart cities* (pp. 1–18). Routledge.

Krisch, N. (2021). The dynamics of international law redux. *Current Legal Problems*, *74*(1), 269–297. https://doi.org/10.1093/clp/cuab008

Kurlantzick, J., & West, J. (2020). *Assessing China's digital road initiative*. Council on Foreign Relations. https://www.cfr.org/china-digital-silk-road/

Lai, C. S., Jia, Y., Dong, Z., Wang, D., Tao, Y., Lai, Q. H., Wong, R. T. K., Zobaa, A. F., Wu, R., & Lai, L. L. (2020). A review of technical standards for smart cities. *Clean Technologies*, *2*(3), 290–310. https://doi.org/10.3390/cleantechnol2030019

Laurent, B., Doganova, L., Gasull, C., & Muniesa, F. (2021). The test bed island: Tech business experimentalism and exception in Singapore. *Science as Culture*, *30*(3), 367–390. https://doi.org/10.1080/09505431.2021.1888909

Lee, H. (2021). The battle to set digital technical standards. *Global Asia*, *16*(4), 42–46.

Lee, T., & Meene, S. van de. (2012). Who teaches and who learns? Policy learning through the C40 cities climate network. *Policy Sciences*, *45*, 199–220. https://doi.org/10.1007/s11077-012-9159-5

Lim, Y., Edelenbos, J., & Gianoli, A. (2023). What is the impact of smart city development? Empirical evidence from a Smart City Impact Index. *Urban Governance*. https://doi.org/10.1016/j.ugj.2023.11.003

Losavio, M. M., Chow, K. P., Koltay, A., & James, J. (2018). The Internet of Things and the Smart City: Legal challenges with digital forensics, privacy, and security. *Security and Privacy*, *1*(3), e23. https://doi.org/10.1002/spy2.23

Lynch, C. R. (2020). Contesting digital futures: Urban politics, alternative economies, and the movement for technological sovereignty in Barcelona. *Antipode*, *52*(3), 660–680. https://doi.org/10.1111/anti.12522

Maier, M., Mcbride, M., & Mcconnell, P. (2018). LinkNYC: Redesigning telecommunication to activate the twenty-first-century creative city. In A. C. D'Almeida (Ed.), *Smarter New York city: How city agencies innovate* (pp. 79–106). Columbia University Press. https://doi.org/10.7312/dalm18374-007

March, H., & Ribera-Fumaz, R. (2018). Barcelona: From corporate smart city to technological sovereignty. In A. Karvonen, F. Cugurullo, & F. Caprotti (Eds.), *Inside smart cities: Place, politics and urban innovation* (pp. 227–242). Routledge.

Marchetti, D., Oliveira, R., & Figueira, A. R. (2019). Are global north smart city models capable to assess Latin American cities? A model and indicators for a new context. *Cities*, *92*, 197–207. https://doi.org/10.1016/j.cities.2019.04.001

Marks, S. (2006). State-centrism, international law, and the anxieties of influence. *Leiden Journal of International Law*, *19*(2), 339–347. https://doi.org/10.1017/S0922156506003335

Marsal-Llacuna, M.-L. (2020). The people's smart city dashboard (PSCD): Delivering on community-led governance with blockchain. *Technological Forecasting and Social Change*, *158*(September), 120150. https://doi.org/10.1016/j.techfore.2020.120150

Mazzocco, I. (2022, July 27). *The AI-surveillance symbiosis in China*. Big Data China. https://bigdatachina.csis.org/the-ai-surveillance-symbiosis-in-china/

Mearsheimer, J. J. (2019). Bound to fail: The rise and fall of the liberal international order. *International Security*, *43*(4), 7–50.

Mena Lora, A. J., Long, J. E., Huang, Y., Baden, L. R., El Sahly, H. M., Follmann, D., Goepfert, P., Gray, G., Grinsztejn, B., Kotloff, K., Rouphael, N., Sobieszczyk, M., Walsh, S. R., Andriesen, J., Shah, K. A., Zhang, Y., Gilbert, P., Janes, H., Gay, C. L., … Kublin, J. G. (2023). Rapid development of an integrated network infrastructure to conduct phase 3 COVID-19 vaccine trials. *JAMA Network Open*, *6*(1), e2251974. https://doi.org/10.1001/jamanetworkopen.2022.51974

Menendez-Blanco, M., & Bjørn, P. (2022). Designing digital participatory budgeting platforms: Urban biking activism in Madrid. In *Computer supported cooperative work: CSCW: An international journal* (Vol. 31, Issue 4). Springer Netherlands. https://doi.org/10.1007/s10606-022-09443-6

Mergel, I., Kleibrink, A., & Sörvik, J. (2018). Open data outcomes: U.S. cities between product and process innovation. *Government Information Quarterly*, *35*(4), 622–632. https://doi.org/10.1016/j.giq.2018.09.004

Mills, D. E., Izadgoshasb, I., & Pudney, S. G. (2021). Smart city collaboration: A review and an agenda for establishing sustainable collaboration. *Sustainability*, *13*(16), 9189. https://doi.org/10.3390/su13169189

Ministry of Digital Development and Information. (n.d.). *Government's personal data protection laws and policies*. Retrieved July 20, 2024, from https://www.mddi.gov.sg/gov-personal-data-protection-laws-and-policies/

Molchanov, M. A., & Molchanova, V. (2022). Smart city—A global project of late capitalism. In M. Lebedeva & V. Morozov (Eds.), *Turning points of world transformation: New trends, challenges and actors* (pp. 27–45). Palgrave Macmillan. https://doi.org/10.1007/978-981-19-1758-5_3

Momen, N. (2023). Privacy and ethics in a smart city: Towards attaining digital sovereignty. In M. Ahmed & P. Haskell-Dowland (Eds.), *Cybersecurity for smart cities: Advanced sciences and technologies for security applications* (pp. 47–60). Springer. https://doi.org/10.1007/978-3-031-24946-4_4

Mookerjee, S. (2021, October 4). *National Smart Coalitions Partnership unites regional efforts*. Route Fifty. https://www.route-fifty.com/management/2021/10/national-smart-coalitions-partnership-unites-regional-efforts/316362/

Naiki, Y. (2024). Smart cities and international trade law. *World Trade Review*, *8*, 1–22. https://doi.org/10.1017/S1474745624000077

National Smart Coalitions Partnership. (n.d.). *About National Smart Coalitions Partnership*. Retrieved July 20, 2024, from https://www.smartcoalitions.com/about

Nelischer, K. (2024). Privately-directed participatory planning: Examining Toronto's Quayside smart city. *Canadian Planning and Policy / Aménagement et Politique Au Canada*, *2024*(1), 10–31. https://doi.org/10.24908/cpp-apc.v2024i1.16781

Noor, E. (2023, March 16). Southeast Asia and the China-US fight for tech supremacy. *AsiaGlobal Online*. https://www.asiaglobalonline.hku.hk/southeast-asia-and-china-us-fight-tech-supremacy

NYC Office of Technology and Innovation. (n.d.). *LinkNYC*. Retrieved July 20, 2024, from https://www.nyc.gov/content/oti/pages/linknyc

Obringer, R., & Nateghi, R. (2021). What makes a city 'smart' in the Anthropocene? A critical review of smart cities under climate change. *Sustainable Cities and Society*, *75*(December), 103278. https://doi.org/10.1016/j.scs.2021.103278

Orakhelashvili, A. (2008). International law and geopolitics: One object, conflicting legitimacies? *Netherlands Yearbook of International Law*, *39*, 155–204. https://doi.org/10.1017/S0167676808001554

Pagani, R., & Fracastoro, G. V. (2021). Smart cities and smart societies: The shock, or the new paradigm for a smart society. In M. Del Giudice & A. Osello (Eds.), *Handbook of research on developing smart cities based on digital twins* (pp. 129–152). IGI Global. https://doi.org/10.4018/978-1-7998-7091-3.ch007

Palacin, V., McDonald, S., Aragón, P., & Nelimarkka, M. (2024). Configurations of digital participatory budgeting. *ACM Transactions on Computer-Human Interaction*, *31*(2). https://doi.org/10.1145/3635144

Papanikolaou, Z., Kefala, F., Karelakis, C., Theodosiou, G., & Goulas, A. (2022). Cities in competition: Is there a link between entrepreneurship and development? *World*, *3*(4), 913–927. https://doi.org/10.3390/world3040051

Perätalo, S., Ahokangas, P., & Iivari, M. (2023). Smart city business model approach: The role of opportunities, values, and advantages. *Innovation: The European Journal of Social Science Research*, 1–25. https://doi.org/10.1080/13511610.2023.2286439

Qandeel, M. (2024). Facial recognition technology: Regulations, rights and the rule of law. *Frontiers in Big Data*, *7*, 1–14. https://doi.org/10.3389/fdata.2024.1354659

Ranchordás, S., & Goanta, C. (2020). The new city regulators: Platform and public values in smart and sharing cities. *Computer Law & Security Review*, *36*(April), 105375. https://doi.org/10.1016/j.clsr.2019.105375

Ranchordás, S., & Klop, A. (2018). Data-driven regulation and governance in smart cities. In V. Mak, E. T. T. Tai, & A. Berlee (Eds.), *Research handbook in data science and law* (pp. 245–273). Edward Elgar Publishing. https://doi.org/10.4337/9781788111300.00018

Reed, M., & Keech, D. (2019). Making the city smart from the grassroots up: The sustainable food networks of Bristol. *City, Culture and Society*, *16*(March), 45–51. https://doi.org/10.1016/j.ccs.2017.07.001

Reuter, T. K. (2020). *Smart city visions and human rights: Do they go together?* (No. 006; Carr Center Discussion Paper Series). https://carrcenter.hks.harvard.edu/files/cchr/files/CCDP_006.pdf

Ricci, L., Di Francesco Maesa, D., Favenza, A., & Ferro, E. (2021). Blockchains for covid-19 contact tracing and vaccine support: A systematic review. *IEEE Access*, *9*, 37936–37950. https://doi.org/10.1109/ACCESS.2021.3063152

Ro, Y., Roh, I., Jahng, J., & Kwon, B. (2023). Determining priority in smart city technologies and services for international development cooperation. *Journal of Computer Information Systems*, *63*(6), 1332–1344. https://doi.org/10.1080/08874417.2022.2150913

Robin, E., & Acuto, M. (2018). Global urban policy and the geopolitics of urban data. *Political Geography*, *66*, 76–87. https://doi.org/10.1016/j.polgeo.2018.08.013

Rowland, Z., Cug, J., & Nica, E. (2022). The geopolitics of smart city digital twins: Urban sensing and immersive virtual technologies, spatio-temporal fusion algorithms, and visualization modeling tools. *Geopolitics, History, and International Relations*, *14*(2), 56–71.

Rühlig, T. (2023). Chinese influence through technical standardization power. *Journal of Contemporary China*, *32*(139), 54–72. https://doi.org/10.1080/10670564.2022.2052439

Ryugen, H., & Akiyama, H. (2020, July 25). *China leads the way on global standards for 5G and beyond*. Nikkei Asia. https://asia.nikkei.com/Politics/International-relations/China-leads-the-way-on-global-standards-for-5G-and-beyond

Sahin, K. (2020, December 18). *The West, China, and AI surveillance*. Atlantic Council. https://www.atlanticcouncil.org/blogs/geotech-cues/the-west-china-and-ai-surveillance/

Samih, H. (2019). Smart cities and internet of things. *Journal of Information Technology Case and Application Research*, *21*(1), 3–12. https://doi.org/10.1080/15228053.2019.1587572

Seoane, M. F. V., & Velasco, C. M. Á. (2024). The Chinese surveillance state in Latin America? Evidence from Argentina and Ecuador. *The Information Society*, *40*(2), 154–167. https://doi.org/10.1080/01972243.2024.2317057

Shadowen, N., Lodato, T., & Loi, D. (2020). Participatory governance in smart cities: Future scenarios and opportunities. In N. Streitz & S. Konomi (Eds.), *Distributed, ambient and pervasive interactions* (pp. 443–463). Springer. https://doi.org/10.1007/978-3-030-50344-4_32

Shah, J., Kothari, J., & Doshi, N. (2019). A survey of smart city infrastructure via case study on New York. *Procedia Computer Science*, *160*, 702–705. https://doi.org/10.1016/j.procs.2019.11.024

Sheehan, M., & Feldgoise, J. (2023, February 27). *What Washington gets wrong about China and technical standards*. Carnegie. https://carnegieendowment.org/research/2023/02/what-washington-gets-wrong-about-china-and-technical-standards?center=global&lang=en

Shi-Kupfer, K., & Ohlberg, M. (2019, April). China's digital rise: Challenges for Europe. *MERICS*. https://merics.org/sites/default/files/2020-06/MPOC_No.7_ChinasDigitalRise_web_final_2.pdf

Smart Surfaces Coalition. (n.d.). *Mission and values*. Retrieved July 20, 2024, from https://smartsurfacescoalition.org/mission-and-values

Söderström, O. (2021). The three modes of existence of the pandemic smart city. *Urban Geography*, *42*(3), 399–407. https://doi.org/10.1080/02723638.2020.1807167

Spicer, Z., & Zwick, A. (2021). A smart city for Toronto: What does Quayside tell us about the state of smart city building? In A. Zwick & Z. Spicer (Eds.), *The platform economy and the smart city* (pp. 245–265). McGill-Queen's University Press. https://doi.org/10.1515/9780228007944-014

Sporchia, F., Marchi, M., Nocentini, E., Marchettini, N., & Pulselli, F. M. (2023). Sub-national scale initiatives for climate change mitigation: Refining the approach to increase the effectiveness of the covenant of mayors. *Sustainability*, *15*(1), 125–139. https://doi.org/10.3390/su15010125

Swiney, C. (2020). The urbanization of international law and international relations: The rising soft power of cities in global governance. *Michigan Journal of International Law*, *41*, 227–278. https://doi.org/10.36642/mjil.41.2.urbanization

Taylor, M. (2022). *China's digital authoritarianism: A governance perspective*. Palgrave Macmillan. https://doi.org/10.1007/978-3-031-11252-2

The White House. (2023). *U.S.-Singapore critical and emerging technology dialogue: Joint vision statement*. https://www.whitehouse.gov/briefing-room/statements-releases/2023/10/12/u-s-singapore-critical-and-emerging-technology-dialogue-joint-vision-statement/

Townsend, A., Popelka, S., & Beroche, H. (2023, May). *The future of urban AI: Global dialogues on urban artificial intelligence*. Cornell Tech Urban Tech Hub. https://bpb-us-w2.wpmucdn.com/sites.coecis.cornell.edu/dist/4/371/files/2023/05/Future-of-Urban-AI-Summary.pdf

Tsai, W.-H., Lin, R., & Wang, H. H. (2023). Urban management in authoritarian China: How the smart city is used to enhance comprehensive law enforcement. *Journal of Contemporary China*. https://doi.org/10.1080/10670564.2023.2251020

U.S.-China Climate-Smart Cities Initiative. (n.d.). Retrieved July 20, 2024, from https://international.lbl.gov/CCWGcities

Ullah, Z., Al-Turjman, F., Mostarda, L., & Gagliardi, R. (2020). Applications of artificial intelligence and machine learning in smart cities. *Computer Communications*, *154*, 313–323. https://doi.org/10.1016/j.comcom.2020.02.069

United Nations. (2019). *World urbanization prospects: The 2018 revision*. https://population.un.org/wup/publications/Files/WUP2018-Report.pdf

US-China Economic and Security Review Commission. (2020, January). *China's smart cities development*. https://www.uscc.gov/sites/default/files/China_Smart_Cities_Development.pdf

US Department of Transportation. ITS Joint Program Office, & Transportation, N. Y. D. of. (2017, March 3). *Connected vehicle pilot deployment program: New York City, New York*. https://rosap.ntl.bts.gov/view/dot/34760

Vanolo, A. (2013). Smartmentality: The smart city as disciplinary strategy. *Urban Studies*, *51*(5), 883–898. https://doi.org/10.1177/0042098013494427

Vinha, L. Da. (2023). Smart for whom? Africa's smart cities and digital authoritarianism. *International Journal of Intelligence and CounterIntelligence*. https://doi.org/10.1080/08850607.2023.2284629

Voorwinden, A. (2021). The privatised city: Technology and public-private partnerships in the smart city. *Law, Innovation and Technology*, *13*(2), 439–463. https://doi.org/10.1080/17579961.2021.1977213

Walters, D. (2011). Smart cities, smart places, smart democracy: Form-based codes, electronic governance and the role of place in making smart cities. *Intelligent Buildings International*, *3*(3), 198–218. https://doi.org/10.1080/17508975.2011.586670

Weber, V. (2023, May). China's smart cities and the future of geopolitics. *LSE IDEAS*. https://www.lse.ac.uk/ideas/Assets/Documents/updates/2023-05-14-SUWeber-SmartCities.pdf

Wiig, A. (2015). IBM's smart city as techno-utopian policy mobility. *City*, *19*(2–3), 258–273. https://doi.org/10.1080/13604813.2015.1016275

Wiig, A. (2016). The empty rhetoric of the smart city: From digital inclusion to economic promotion in Philadelphia. *Urban Geography*, *37*(4), 535–553. https://doi.org/10.1080/02723638.2015.1065686

Wilson, B., & Chakraborty, A. (2019). Planning smart(er) cities: The promise of civic technology. *Journal of Urban Technology*, *26*(4), 29–51. https://doi.org/10.1080/10630732.2019.1631097

Xie, S., Luo, N., & Yarime, M. (2024). Data governance for smart cities in China: The case of Shenzhen. *Policy Design and Practice*, *7*(1), 66–86. https://doi.org/10.1080/25741292.2023.2297445

Xu, Y., Cugurullo, F., Zhang, H., Gaio, A., & Zhang, W. (2024). The emergence of artificial intelligence in anticipatory urban governance: Multi-scalar evidence of China's transition to city brains. *Journal of Urban Technology*. 1–25. https://doi.org/10.1080/10630732.2023.2292823

Yang, F., & Xu, J. (2018). Privacy concerns in China's smart city campaign: The deficit of China's Cybersecurity Law. *Aisa & The Pacific Policy Studies*, *5*(3), 533–543. https://doi.org/10.1002/app5.246

Yang, R., & Zhen, F. (2024). Smart city development models: A cross-cultural regional analysis from theory to practice. *Research in Globalization*, *8*(November 2023), 100221. https://doi.org/10.1016/j.resglo.2024.100221

Zarakol, A. (2017). Theorising hierarchies: An introduction. In A. Zarakol (Ed.), *Hierarchies in world politics* (pp. 1–14). Cambridge University Press. https://doi.org/10.1017/9781108241588.002

Zhang, C., Zhang, M., & Xiao, C. (2022). From traditional infrastructure to new infrastructure: A new focus of China's Belt and Road Initiative diplomacy? *Eurasian Geography and Economics*, *63*(3), 424–443. https://doi.org/10.1080/15387216.2022.2039740

Zhang, J., Hua, X.-S., Huang, J., Shen, X., Chen, J., Zhou, Q., Fu, Z., & Zhao, Y. (2019). City brain: Practice of large-scale artificial intelligence in the real world. *IET Smart Cities*, *1*(1), 1–37. https://doi.org/10.1049/iet-smc.2019.0034

Zúñiga, N., Burton, S. D., Blancato, F., & Carr, M. (2024). The geopolitics of technology standards: Historical context for US, EU and Chinese approaches. *International Affairs*, *100*(4), 1635–1652. https://doi.org/10.1093/ia/iiae124

10

NAVIGATING INTERNATIONAL LEGAL FRONTIERS FOR SUSTAINABLE URBANIZATION

Leila Hennaoui

Introduction

The accelerating pace of global urbanization has fundamentally transformed the landscape of international affairs. Cities, wielding ever-increasing influence, are no longer mere components of nation-states. This shift coincides with the emergence of smart cities, a new paradigm characterized by advanced technological infrastructure and innovative governance systems. Smart cities are progressively recognized as pivotal agents bridging local and global concerns, exemplified by efforts to align sustainable development with urban management (Gauthier, 2009). This growing attention necessitates a fundamental reevaluation of the role of cities in international law and governance. However, classical international law remains rooted in a state-centric approach, viewing cities as subordinate entities lacking autonomy in the international sphere (Sossai, 2021). This disconnect between the growing influence of smart cities and their legal status necessitates a critical reevaluation of the role of cities in international law and governance.

Here, I argue that smart cities, with their unique capabilities, represent a distinctive urban development paradigm with the potential to augment a city's international agency and impact on global issues, particularly sustainability. This enhanced agency and effect is achieved through two key mechanisms. The first mechanism involves leveraging transnational city networks. These networks function as international platforms fostering collaboration through knowledge sharing and joint projects (Durmus & Oomen, 2022; Picavet et al., 2023; Triviño-Salazar, 2023). Smart cities can utilize these platforms to tackle pressing sustainability challenges. For instance, collaboration can lead to advancements in climate change mitigation through smart grids

DOI: 10.4324/9781003439325-14

and resource optimization through data-driven management systems. The second mechanism lies in the technological core of smart cities. This emphasis on innovation and advanced technology opens up new avenues for pursuing sustainability objectives (Vodák, Šulyová, & Kubina, 2021; Carro-Suárez, Sarmiento-Paredes, & Nava, 2023). Smart city technologies can be harnessed to develop and implement solutions for clean energy production, waste management, and resource conservation.

This chapter aims to illuminate smart cities' evolving position as normative actors who can shape and influence international norms (Durmus & Oomen, 2022), particularly those related to sustainability. This influence is achieved by harnessing their international character and through several tools, including transnational city networks, advocacy efforts, and the utilization of soft law instruments. By acting in this capacity, smart cities have the potential to become catalysts for translating global aspirations into concrete actions that benefit both their local inhabitants and the broader international community in achieving global sustainability objectives.

This chapter unfolds in a structured manner, embarking on a discursive analysis of smart cities' international character in Section "Discursive analysis of smart cities international character". It delves into the evolving role of smart cities within international law, portraying them as implementers challenging conventional perspectives. Transitioning further, Section "Smart cities: shaping and implementing global norms for sustainability" delves into the active role of smart cities as global actors, engaging in international forums, shaping norms through soft law tools, and intricately interweaving technology, governance, sustainability, and resilience. Following this, Section "Operationalizing sustainability through smart cities: advancing the UN's sustainable development goals" explores the operationalization of sustainability through smart cities, specifically focusing on their contributions to advancing the United Nations (UN's) Sustainable Development Goals (SDGs). This includes the integration of SDGs into smart city development and the adoption of human-centric models. As the chapter unfolds, each section contributes to shedding light on the multifaceted contributions of smart cities in shaping a sustainable, interconnected world. The concluding section emphasizes the pivotal role of a robust legal framework in shaping the future trajectory of inclusive and humanized smart cities, thereby contributing significantly to global sustainability objectives.

Discursive analysis of smart cities' international character

The urban landscape is undergoing a significant transformation. The concept of "smart cities" has emerged as a dominant theme, supplanting "sustainable city" as the central focus in discussions about urban development (Joss et al., 2019; Moir, Moonen, & Clark, 2014). This shift reflects a broader global discourse centered on technological advancements shaping urban development strategies. Notably, cities are actively involved in an internationalization process, extending beyond local boundaries to strategically position themselves within the global arena by aligning their internationalization efforts with the broader discourse on smart urban development (Marchesani et al., 2021). In this evolving context, cities, as dynamic and discursive practices, are continuously shaped by global norms and stakeholder strategies (Marčenko, 2021).

Accordingly understanding the international character of smart cities requires delving deeper than just technological advancements. We need to examine how stakeholders frame these entities within the global discourse network. This is where discursive analysis emerges as a potent tool. Discursive analysis examines how language is used by various stakeholders, including governments, multinational corporations, and international organizations, to shape the discourse around smart cities and influence the development of global norms. By analyzing this global network of communication, we aim to illuminate the collective understanding of smart cities as interconnected actors on a global scale.

Global collaboration and the international character of smart cities

The evolving landscape of smart cities transcends a purely technological focus. While advancements in digital technologies remain foundational, the discourse surrounding smart urban development increasingly prioritizes strategies for global engagement (Joss et al., 2019). This shift reflects a growing recognition of the international dimension's centrality to smart city initiatives. Cities actively align their internationalization efforts with the broader discourse on smart cities, fostering a nexus between international management strategies and the evolution of successful urban development models (Marchesani et al., 2021). This emphasis on global collaboration underscores the critical role of international perspectives in shaping and justifying smart city programs.

To navigate this complex international landscape, cities actively engage in collaborative efforts to advance smart technologies, emphasizing international partnerships (Kihlgren Grandi, 2020; Marchesani et al., 2021). Major global city networks, including the United Cities and Local Governments (UCLG), the Local Governments for Sustainability (ICLEI),[1] and the World Smart Sustainable Cities Organization (WeGO), align with the UN vision, integrating technologies to promote sustainable development (Kihlgren Grandi, 2020). This collaborative approach focus is on the development of global norms and policy standards to accelerate best practices and enhance public trust in smart city technologies (Kihlgren Grandi, 2020; Marchesani et al., 2021).

Within this frame discourse, the UCLG plays a pivotal role in internationalizing the discourse surrounding smart cities. As a global advocate for cities, UCLG hosts a dedicated Digital Cities Community of Practice that actively fosters collaboration among local authorities on a global scale, contributing to the global interconnectedness of smart city narratives (Smart Cities Study, 2019). The annual Smart City Expo World Congress, co-organized by UCLG, exemplifies the inherently international character of smart cities, fostering knowledge exchange, collaboration, and cross-border problem-solving. This global forum serves as a melting pot where smart cities globally learn, adapt, and contribute to a shared vision of interconnected progress (Smart Cities Study, 2019).

National governments in the international smart city landscape

National governments play a crucial role in enabling the international dimension of smart city initiatives. Due to jurisdictional limitations, city-level implementation hinges on the consent and actions of national authorities. This is particularly showcased in many national smart city strategies, in Spain, India, Canada, and Rwanda (Kihlgren Grandi, 2020). These national strategies often serve as frameworks for international cooperation. Developing countries, in particular, leverage such frameworks to access vital subventions and technical support from international organizations and development banks and also from developed countries (Kihlgren Grandi, 2020). National governments further strengthen the international interconnectedness of smart cities by fostering international collaboration (New, Castro, & Beckwith, 2017).

A comprehensive analysis of government discourse on smart cities, drawn from systematically examined primary and secondary sources encompassing the United Kingdom, Italy, China, and India (The UK Department for Business, Innovation, & Skills, 2013 ; Dameri et al., 2019 ; Purandare, 2023), provides synthesized insights into the global landscape of urban development strategies. This analysis illuminates a shared international orientation among local governments in shaping smart city narratives.

For instance, a 2013 publication by the UK Department for Business, Innovation & Skills reflects early recognition of the global challenges confronting cities worldwide but also emphasizes the instrumental role of smart city concepts in addressing them. Importantly, the document articulates a strategic vision that extends beyond national borders, recognizing the burgeoning global market for smart urban systems. This government paper substantiates the argument that local and national governments are not only cognizant of but actively embracing the international discourse on smart cities.

Moreover analyses of smart city initiatives across different countries further solidify the concept's international character. Comparison of initiatives in Italy and China, as examined by Dameri et al. (2019), underscores the glocal nature of smart city strategies, highlighting the interconnectedness between global trends, such as the emphasis on data and technology, and local implementations at the municipal level, where factors like cultural nuances and existing infrastructure play a significant role (Dameri et al., 2019).

This glocal nature is further evident in the convergence with global rhetoric. Uttara Purandare's (2023) discourse analysis of India's Smart Cities Mission exemplifies this convergence. It reveals that the mission aligns with global trends in several ways: as it prioritizes data-driven solutions and technological advancements to address urban challenges. This aligns with the broader global goals of fostering economic growth alongside sustainable development in smart cities. Interestingly, the mission acknowledges recent critiques of purely technology-driven approaches. This resonates with a global shift toward more holistic approach (Purandare, 2023).

Smart cities: Shaping and implementing global norms for sustainability

The complex network of smart cities navigating global frontiers unfolds through strategic diplomacy, international collaborations, and the use of soft law instruments, shaping a set of sustainable urban development norms. This section delves into the multifaceted strategies employed by smart cities and their pivotal role in shaping and implementing global norms for sustainability.

City diplomacy in smart cities

In the contemporary discourse on smart cities, the imperative for sustainable urban development has become paramount. One prominent avenue through which smart cities contribute to the global sustainability agenda is city diplomacy. City diplomacy refers to the process by which cities engage with international actors like organizations, governments, and other cities, aiming to represent their interests and collaborate on global challenges (Van der Pluijm & Melissen, 2007). Characterized by its soft power approach, it emerges as a transformative force in international politics, pushing for human-centric solutions to global challenges. Accordingly smart cities actively participate in negotiations, forming networks, and engaging with foreign counterparts (Bolewski, 2020).

City diplomats, guided by mayoral leadership, emerge as key international actors, actively contributing in shaping global norms through a spectrum of collaborations. From North-South partnerships to participation in regional smart city networks and contributions to global alliances; these collaborations facilitate knowledge transfer, introduce replicable technologies, and define policy standards, contributing to a collective effort in accelerating best sustainability practices worldwide (Kihlgren Grandi, 2020).

Inspired by the strategic integration of social responsibility into corporate diplomacy, which enhances public perception and legitimizes corporations' societal roles, Bolewski (2020) suggests that smart cities can leverage a similar approach. By positioning themselves as key partners in addressing global challenges and demonstrating their unique capabilities in data-driven decision-making and technological innovation, smart cities can further strengthen their contribution – and justify their rightful place – in shaping global sustainability norms and practices (Bolewski, 2020).

Global collaborations and international alliances

Smart cities' contribution to the establishment of global norms is consecrated by their engagement through global partnerships like WeGO, C40 Cities Climate Leadership Group, and the Group of Twente (G20) Global Smart Cities Alliance. These partnerships contribute to shaping global norms by addressing climate issues, fostering data-driven measures, and defining international standards for technology governance (Kihlgren Grandi, 2020). The G20 Global Smart Cities Alliance emerges as exemplary, demonstrating the effectiveness of urban networks in promoting sustainable urban development on a global scale (Harakan, 2023).

The G20 Global Smart Cities Alliance on Technology Governance was launched in 2019. Managed by the World Economic Forum, it aims to define global norms and policy standards and fosters collaboration among city networks such as UCLG , ICLEI , and WeGO (Kihlgren Grandi, 2020; Buchoud, 2020). Bringing together cities, businesses, and organizations, the alliance aims to promote the adoption of smart city solutions that enhance sustainability, resilience, and quality of life for urban residents. This further showcases the G20's overarching engagement, which effectively establishes a critical connection between the global economic landscape and urban challenges. This linkage implies that solutions geared toward urban issues, including the implementation of smart cities, carry a consequential global impact, particularly in fostering sustainability, as the G20 commitment to fostering an innovative, interconnected, and inclusive world economy places sustainability at the forefront (Buchoud, 2020). This stands as an exemplary illustration of smart cities' international networks and collaborative efforts that transcend geographical boundaries, uniting cities in a shared quest for sustainable and equitable urban development (Harakan, 2023).

Use of soft law instruments for sustainability

In the dynamic landscape of smart cities, effectively navigating the global challenge of shaping sustainability norms and policies requires strategic tools that transcend traditional legal frameworks. Soft law instruments emerge as the perfect companions for smart cities on the international stage, offering a flexible and collaborative approach to tackle complex sustainability issues (Voorwinden & Ranchordas, 2021). Accordingly smart cities are forging international networks and engaging with standardization bodies.

Transnational smart city networks focused on climate change and sustainability, such as the above-mentioned partnerships and alliances, have emerged as integral soft law tools (Voorwinden & Ranchordas, 2021). These networks, exemplified by their capacity to consolidate resources and best practices, operate as a means to bypass state-centric hierarchies. They underscore the importance of sharing opportunities for local action (Acuto, 2013).

Integral to the global smart city use of soft law tools is the adoption of standards and performance indicators, developed by international standardization bodies such as the International Organization for Standardization (ISO), the International Telecommunication Union (ITU), and the International Electrotechnical Commission (IEC); together they create a framework that underscores global collaboration in shaping sustainable norms (Estevez et al., 2021). In addition to these standard-setting organizations, dedicated study groups and committees on smart and sustainable cities contribute to the development of sustainability indicators (Voorwinden & Ranchordas, 2021). The emerging framework of international standards illuminates shared dimensions such as education, environment, energy, health, safety, and governance underscoring a collective commitment to addressing fundamental aspects of sustainability (Estevez et al., 2021).

Soft law, embodied in these transnational city networks and these international standardization bodies, establishes a common language and a robust evaluation framework laying the foundation for measurable targets and shared best practices in the realm of sustainable urban development (Voorwinden & Ranchordas, 2021). While lacking legally binding force, these instruments help attain legitimacy through their consensus-driven processes (Brunsson et al., 2012). The impact of these soft law instruments, while yet to be fully determined, provides a foundation for establishing a common ground in the pursuit of sustainable urban development (Voorwinden & Ranchordas, 2021; Huovila et al., 2019).

Operationalizing sustainability through smart cities: Advancing the UN's Sustainable Development Goals

Cities are increasingly recognized as essential hubs for bridging local and global concerns (Eslava, 2015). In this context, smart cities, with their advanced technological infrastructure and efficient governance systems, emerge as ideal candidates for navigating international legal frontiers in implementing sustainable urbanization. Sustainability, as an overarching goal, encompasses a broad vision of creating resilient, eco-friendly, and socially inclusive urban environments. However, within the realm of sustainable urbanization, it is primordial to focus on a well-defined roadmap – the internationally agreed-upon SDGs of the 2030 Agenda. SDGs provide a structured framework, delineating precise targets for cities to achieve in their quest for a more sustainable future. In that context, cities are deemed a "privileged international locus" for sustainable development (Porras, 2009) and are increasingly recognized in UN frameworks and the aforementioned SDGs (Kosovac & Acuto, 2020).

This section delves into the multifaceted contributions of smart cities to both sustainability in its broader sense and the precise implementation of SDGs. It underscores the need for a comprehensive and inclusive approach beyond technological tools, recognizing the dual responsibility of smart cities to contribute to overarching sustainability aspirations and adhere to the specific targets outlined in the SDGs. The exploration navigates through the integration of SDGs into smart city development, and the human-centric models.

This nuanced approach allows us to appreciate the holistic vision of sustainability while acknowledging the need for targeted actions embodied in the SDGs, ultimately paving the way for smart cities to become dynamic contributors to global sustainability objectives

Integrating SDGs into smart city development

Smart cities, particularly with their emphasis on technological innovation, have the inherent potential to advance specific SDGs in the urban context. As Yang and Rajabifard (2019) rightly noted, certain SDGs are inherently more attainable within urban environments than others.

Smart cities actively contribute to achieving SDG 8 (Decent Work and Economic Growth) by embracing circular economy and green economy models via advanced technologies that facilitate financial inclusion for businesses, especially through blockchain-based platforms (Sharifi et al., 2024). They uniquely also enable the achievement of SDG 9 (Industry, Innovation, and Infrastructure) by serving as platforms for diverse and advanced innovations, fostering economic growth across various sectors, including industry and infrastructure (Ismagilova et al., 2022). Particularly, blockchain technology in smart cities facilitates green production, sustainable supply chain logistics, and innovative finance models, enhancing economic growth (Sharifi et al., 2024). Powered by advanced technologies like blockchain, Internet of Things (IoT) and Information and Communication Technologies (ICT), these cities emerge as pivotal hubs for achieving SDG 12 (Responsible Consumption and Production). These technologies optimize resource utilization. Circular economy models, facilitated by AIoT, play a crucial role in applications such as waste water management, energy-efficient smart grids, and shared resource utilization. Blockchain technology becomes instrumental in monitoring pollution, overseeing groundwater, and ensuring compliance with sustainability standards through product traceability (Parmentola et al., 2022; Sharifi et al., 2024).

Smart cities' contribution to attainment of SDG 11 (Sustainable Cities and Communities) is unparalleled. Leveraging advanced technologies like big data analytics, drones, blockchain, and AI-enabled monitoring, they optimize urban management, enhance resilience, and address challenges such as air pollution and resource management. The integration of smart solutions in shared mobility, modal shifts, and urban accessibility aligns with SDG 11's objectives, fostering sustainable and efficient urban development (Sharifi et al., 2024).

However, it is crucial to acknowledge the potential challenges and trade-offs; concerns such as privacy and cybersecurity, infrastructure costs, increased cost of living, digital divide and skill gaps, biased decision-making, reproduction of social biases, and limited legal frameworks can pose significant hurdles. These challenges underscore the complexity of adopting a technology-based approach in isolation (Sharifi et al., 2024; Yang & Rajabifard, 2019; Inghirami, 2022).

Human-centric models and inclusive governance

It is evident that a holistic approach, encompassing not only technological advancements but also robust regulatory frameworks, skill development, and considerations for societal well-being, is imperative (Yang & Rajabifard, 2019; Inghirami, 2022). Striking a balance between technological innovation and addressing these challenges will be pivotal in ensuring the sustainable and inclusive development of smart cities aligned with the broader SDGs.

To address criticisms and align smart city development with SDGs, there is a call for the integration of a human-centric model, considering the impact on human rights, and social justice (Inghirami, 2022). Thus in order to operationalize the SDGs, smart cities need to implement these goals' underlying fundamental rights as they are often better equipped to achieve them than states (Advisory Council on International Affairs, 2019).

The intertwining of SDGs with fundamental human rights principles emerges as a guiding framework for implementing those goals within smart cities. SDG11 is the principal point of departure that serves as a blueprint for smart cities' SDG achievement. This goal extends beyond mere technological integration, calling for pollution reduction, affordable housing, and the preservation of cultural and natural heritage (Inghirami, 2022).

To truly achieve these goals, a commitment to leaving no one behind is essential, aligning with human rights principles of equality and non-discrimination. In the smart city context, where technology often intersects with issues of privilege, it becomes imperative to ensure accessibility, inclusivity, and equity for all residents. This resonates with the broader objectives of SDGs and human rights frameworks, offering a transformative vision where smart cities become not only technologically advanced but socially and environmentally sustainable havens for diverse communities (Inghirami, 2022).

The "right to the city" emerges also as a transformative force, advocating for inclusive urban governance and participatory democracy, aligning with the pursuit of social justice and inclusive urbanism (Inghirami, 2022). Originating from Lefebvre's call for collective urban rights, it challenges the prevailing urban development model that prioritizes privatization and economic interests. In the context of smart cities, both human rights and SDG 11 are redefining the right to the city. This resignification guided by human rights in the face of emerging digital rights, technologies, and socio-environmental transformations is a vital condition for the reorganization of urban spaces to enhance people's quality of life in a "hypercomplex society" (Ferraresi, 2023). Accordingly the "right to the city" envisions a bottom-up approach that involves local perspectives and active community participation in decision-making processes and shaping urban policies. By emphasizing public spaces that encourage sociability, culture, and the inclusion of diverse identities, the "right to the city" becomes an integral framework for developing smart cities that are not only technologically advanced but also socially and culturally responsive (Inghirami, 2022).

Conclusion

Cities' international character and influence have become the focal point of ongoing debates. In this chapter, we assert that this global character is significantly heightened and accentuated when cities embrace smart technologies. The influence and impact of smart cities emerge as even more potent in their pursuit of sustainability. Guided by this hypothesis, our journey navigated through the intricate web of international networks and engagements that smart cities establish, particularly in their quest for sustainability. Whether employing soft law mechanisms to shape international norms or utilizing their technological advancements to implement the SDGs, our findings underscore the pivotal role of smart cities in the global landscape.

Our exploration revealed that the discourse on smart cities' international character is intricately linked with their strategies for global engagement. They strategically position themselves in the global arena, aligning internationalization efforts with the broader discourse on smart urban development. International collaboration is evident in efforts such as WeGO and partnerships with major global city networks. Furthermore, the study illuminated the adoption of soft law instruments, such as standards and performance indicators developed by international bodies, forming a collaborative framework for sustainable urban development.

Consequently, we assert a compelling need to formally recognize the international dimension of smart cities and fostering their transformative role. For that we propose a two-pronged approach.

Expanding the legal framework: While soft law instruments have played a valuable role, formalizing certain norms through multilateral agreements can provide greater stability and enforceability, especially when dealing with powerful technology giants. This could involve agreements on data privacy, ethical AI development, or sustainable infrastructure standards.

Empowering smart cities: National governments should grant smart cities greater international visibility and autonomy to engage directly with international organizations, nongovernmental organizations (NGOs), and other relevant stakeholders. This could involve formalizing channels for collaboration, adopting soft law norms developed by city networks, and providing resources for capacity building and international outreach.

In conclusion, recognizing and nurturing the international dimension of smart cities, while emphasizing the importance of a robust legal framework, can pave the way for inclusive and humanized smart cities. Consequently, this combined approach unlocks their full potential as drivers of a more sustainable and equitable future for all. This transformative potential of smart cities in the international landscape underscores the need to recalibrate legal frameworks to accord them the recognition and agency they deserve in shaping and implementing sustainability norms on the global stage.

Note

1 ICLEI – Local Governments for Sustainability – was founded in 1990 as the International Council for Local Environmental Initiatives.

References

Acuto, M. (2013). *Global cities, governance and diplomacy*. London: Routledge.

Advisory Council on International Affairs. (2019, May). *Sustainable development goals and human rights: An indivisible bond* (No. 110). https://www.advisorycouncilinternationalaffairs.nl/documents/publications/2019/05/10/sustainable-development-goals-and-human-rights

Bolewski, W. (2020, October 8). Smart City diplomacy inspired by corporate diplomacy. Keynote prepared for the "DevOps Competences for Smart Cities" Workshop at the "All Digit Summit 2020" in Berlin.

Brunsson, N., et al. (2012). The dynamics of standardisation: Three perspectives on standards in organisation studies. *Organization Studies*, *33*, 613–632.

Buchoud, N. J. A. (2020). Making the case for G20 action on urbanization. *Global Solutions Journal*. https://www.global-solutions-initiative.org/press-news/g20-action-urbanization-nicolas-buchoud/

Carro-Suárez, J., Sarmiento-Paredes, S., & Nava, D. (2023). Smart and sustainable cities: A new urban transformation. In *Sustainable regional planning*. IntechOpen. doi:10.5772/intechopen.110234

Dameri, R. P., Benevolo, C., Veglianti, E., & Li, Y. (2019). Understanding smart cities as a glocal strategy: A comparison between Italy and China. *Technological Forecasting and Social Change*, *142*, 26–41. https://doi.org/10.1016/j.techfore.2018.07.025

Durmus, E., & Oomen, B. (2022). Transnational city networks and their contributions to norm-generation in international law: The case of migration. *Local Government Studies*, *48*(6), 1048–1069. https://doi.org/10.1080/03003930.2021.1932478

Eslava, L. (2015). *Local space, global life: The everyday operation of international law and development*. New York: Cambridge University Press.

Estevez, E., Cenci, K., Fillottrani, P. R., & Janowski, T. (2021). Review of international standards and policy guidelines for smart sustainable cities. In E. Estevez, T. A. Pardo, & H. J. Scholl (Eds.), *Smart cities and smart governance, towards the 22nd century sustainable city* (pp. 69–99). Springer. https://doi.org/10.1007/978-3-030-61033-3_4

Ferraresi, C. S. (2023). The right to the (smart) city, human rights and 2030 agenda: Smart cities and the cities (r)evolution. In I. C. Fonseca (Ed.), *Smart cities and law, e.governance and rights*. Wolters Kluwer Italia.

Gauthier, M. (2009). Urban planning and sustainable development. Environnement urbain. *Urban Environment*, *3*. http://journals.openedition.org/eue/894

Harakan, A. (2023). Indonesia's city diplomacy in promoting global sustainability context through urban networks. *Ganaya: Jurnal Ilmu Sosial Dan Humaniora*, *6*(3), 529–538. https://doi.org/10.37329/ganaya.v6i3.2519

Huovila, A., et al. (2019). Comparative analysis of standardized indicators for Smart sustainable cities: What indicators and standards to use and when? *Cities*, *89*, 141–153.

Inghirami, P. (2022). Towards a human smart sustainable city: The case of Barcelona. Master's thesis, MSc in Law, Digital Innovation and Sustainability, Luiss Guido Carli. http://tesi.luiss.it/34688/

Ismagilova, E., Hughes, L., Rana, N. P., & Dwivedi, Y. K. (2022). Security, privacy and risks within smart cities: Literature review and development of a Smart City Interaction Framework. *Information Systems Frontiers*, *24*, 393–414. https://doi.org/10.1007/s10796-020-10044-1

Joss, S., Sengers, F., Schraven, D., Caprotti, F., & Dayot, Y. (2019). The smart city as global discourse: Storylines and critical junctures across 27 cities. *Journal of Urban Technology*, *26*(1), 3–34. DOI: 10.1080/10630732.2018.1558387

Kihlgren Grandi, L. (2020). *City diplomacy. Cities and the global politics of the environment*. Palgrave Macmillan Cham. https://doi.org/10.1007/978-3-030-60717-3

Kosovac, A., & Acuto, M. (2020). Acknowledging urbanization: A survey of the role of cities in UN frameworks. *Global Policy*, *11*(3), 293–304.

Marčenko, M. (2021). The role of the city as a discursive practice in international law and governance (Doctoral dissertation, Universiteit van Amsterdam, Faculty of Law [FdR]). https://pure.uva.nl/ws/files/63630020/Thesis.pdf

Marchesani, F., Iaia, L., Masciarelli, F., & Christofi, M. (2021, October). Smart city's Internationalization and International Management Strategies in the digital era: A systematic literature review. Paper presented at *ITAIS 2021 - Digital resilience and sustainability: people, organizations, and society*, University of Trento, Italy.

Moir, E., Moonen, T., & Clark, G. (2014). *What are future cities? Origins, meanings, uses. london: future cities catapult. foresight future of cities project.* London: Univ. of Westminster.

New, J., Castro, D., & Beckwith, M. (2017, October 30). *How national governments can help smart cities succeed.* The Center for Data Innovation.

Parmentola, A., Petrillo, A., Tutore, I., & De Felice, F. (2022). Is blockchain able to enhance environmental sustainability? A systematic review and research agenda from the perspective of sustainable development goals (SDGs). *Business Strategy and the Environment*, *31*(1), 194–217. https://doi.org/10.1002/bse.2882

Picavet, M.E.B., de Macedo, L.S.V., Bellezoni, R.A. et al. (2023). How can Transnational Municipal Networks foster local collaborative governance regimes for environmental management? *Environmental Management*, *71*, 505–522. https://doi.org/10.1007/s00267-022-01685-w

Porras, L. M. (2009). The city and international law: In pursuit of sustainable development. *Fordham Urban Law Journal*, *36*(3), 537. https://ir.lawnet.fordham.edu/ulj/vol36/iss3/7

Purandare, U. (2023). The language of smart cities. *South Asia Multidisciplinary Academic Journal*, *30*. http://journals.openedition.org/samaj/8706 DOI: https://doi.org/10.4000/samaj.8706

Sharifi, A., Allam, Z., Bibri, S. E., & Khavarian-Garmsir, A. R. (2024). Smart cities and sustainable development goals (SDGs): A systematic literature review of co-benefits and trade-offs. *Cities*, *146*,. https://doi.org/10.1016/j.cities.2023.104659

Smart Cities Study. (2019). *International study on the situation and future trends in smart governance* (3rd ed.). UCLG Community of Practice on Digital Cities. https://www.uclg.org/sites/default/files/uclg_smartcitiesstudy2019_digital_en_0.pdf

Sossai, M. (2021). Invisibility of cities in classical international law. In H. P. Aust, J. E. Nijman, & M. Marcenko (Eds.), *Research handbook on international law and cities*. Cheltenham, UK and Northampton, MA: Edward Elgar Publishing

The UK Department for Business, Innovation & Skills. (2013, October). Smart cities: Background paper. https://www.gov.uk/government/publications/smart-cities-background-paper

Triviño-Salazar, J. C. (2023). Transnational city networks on migration and integration and local collaborative governance: Establishing the Nexus. *International Migration Review*, 57(4), 1402–1426. https://doi.org/10.1177/01979183231154558

Van der Pluijm, R., & Melissen, J. (2007, April). *City diplomacy: The expanding role of cities in international politics*. Netherlands Institute of International Relations Clingendael.

Vodák, J., Šulyová, D., & Kubina, M. (2021). Advanced technologies and their use in smart city management. *Sustainability*, *13*(10), 5746. https://doi.org/10.3390/su13105746

Voorwinden, A., & Ranchordas, S. (2021). Soft law in city regulation and governance. In U. Morth, E. Korhea-aho, M. Eliantonio (Eds.), *Edward Elgar research handbook on soft law*. University of Groningen Faculty of Law Research Paper No. 19/2021. SSRN: https://ssrn.com/abstract=3978959

Yang, Z., & Rajabifard, A. (2019). Implementing SDGs in smart cities beyond digital tools. In A. Rajabifard (Ed.), *Sustainable development goals connectivity dilemma: Land and geospatial information for urban and rural resilience* (pp. 175–197). CRC Press. https://doi.org/10.1201/9780429290626-11

11

INDIVIDUALS' PRIVACY AND DATA COLLECTION IN THE CONTEXT OF SMART CITIES

Law and policy review

Ridoan Karim

Introduction

Smart cities have become a significant focal point of discussion. They essentially encompass the utilization of technology-driven solutions to enhance citizens' lives, improve government interaction, and promote sustainable development (Ismagilova et al., 2020; Yahia et al., 2021; Yu & Xu, 2018). The rapid expansion of these intelligent urban environments and the extensive collection of urban data underscore the growing imperative to safeguard individuals' privacy. While utilizing data offers various advantages, such as enhancing city planning, optimizing resource allocation, and improving service delivery (Hashem et al., 2016), it also raises concerns about privacy violations and the potential misuse of personal information (Sookhak et al., 2018). For example, there remain limited opportunities in smart cities for individuals to provide meaningful consent to the processing of their personal data (Edwards, 2016). Other challenges involve the extent to which smart cities gather private data from routine public interactions, the "privatization" of both infrastructure and data ownership, the repurposing of "big data" derived from the Internet of Things (IoT) in smart cities, and the storage of such data in the Cloud (Edwards, 2016).

In light of these challenges, it is crucial to strike a delicate balance between harnessing the benefits of data for societal progress and ensuring the protection of individuals' privacy. Many researchers are now exploring avenues to achieve the balance through the study of technology (Lai et al., 2020), urban and environment (Huang et al., 2021), and sociology (Kitchin, 2016; de Gracia-Soriano et al., 2021). Inevitably, the rise of smart cities raises significant concerns about personal privacy. Smart cities rely on extensive data

DOI: 10.4324/9781003439325-15

collection through sensors, cameras, and other IoT devices, which can lead to continuous monitoring of citizens' activities (Syed et al., 2021). There is often insufficient transparency about what data is being collected, how it is used, and who has access to it (Brauneis & Goodman, 2018). This lack of clarity can undermine trust and leave citizens unsure about their privacy rights (Brauneis & Goodman, 2018). Additionally, the vast amounts of data collected in smart cities need to be securely stored and transmitted (Syed et al., 2021). Vulnerabilities in these systems can lead to data breaches and unauthorized access to sensitive personal information (Syed et al., 2021).

Existing legal frameworks may not be sufficient to address the unique privacy challenges posed by smart cities, leading to gaps in protection and enforcement. This concern highlights the urgent need for strong policy and legal measures to address the above-mentioned challenges. However, the question remains: what are the legal and policy measures to be incorporated to address the privacy and data protection challenges posed by smart cities? To answer such a question, this chapter aims to analyze existing legal principles to identify potential issues and gaps in the current legal framework. The chapter emphasizes the importance of addressing a global challenge by identifying potential solutions and evaluating their practicality and effectiveness. By doing so, we can establish the groundwork for developing laws and regulations that protect privacy while also reaping the benefits of smart cities.

The chapter begins by exploring the multifaceted relationship between privacy, social principles, and the evolution of smart cities in section 2. This section sets the stage for further legal and policy analysis by discussing the integration of various technologies into urban infrastructure. Following this, a detailed discussion of the policy debates relating to security versus privacy is presented relating to smart cities in section 3. This section also addresses the broader social and ethical implications of smart cities, focusing on issues such as inequality, privacy, and democratic governance. In section 4, I explore the current legal challenges and propose potential solutions to balance technological advancements with individual rights. Finally, in section 5, I conclude the chapter by summarizing the key arguments. Through these sections, I aim to provide a comprehensive understanding of the critical issues at the intersection of technology, law, and society in the context of smart cities.

Privacy, social principles, and smart cities

Over the last three decades, the concept of smart cities has undergone a significant transformation, shaped by advancements in internet working technology. At the core of this evolution lies the IoT, envisioning the seamless integration of interconnected network technology (Yang & Xu, 2018; Miorandi et al., 2012). This integration harmoniously blends conventional technologies like microwaves, security systems, and cars with emerging technologies such as personal health tech and smartphones (Chavhan, 2022).

The IoT acts as the conceptual precursor to contemporary smart city policies, originating from IBM's "Smart Earth" vision as part of its "Smarter Planet" initiative in November 2008 (Yang & Xu, 2018; Hu & Li, 2012). The fundamental idea involves embedding information technology systems into physical objects to enhance urban experiences for citizens (Yang & Xu, 2018). The city's "smartness" is closely tied to its reliance on networking technologies, facilitating communication within urban spaces.

The term "smart city" carries diverse definitions and connotations, often interwoven with other technical language describing the contemporary urban context, including "intelligent", "innovative", "networked", "wired", "digital", "creative", and "cultural" (Nam & Pardo, 2011). This linguistic diversity, combined with interdisciplinary perspectives, complicates efforts to derive a clear meaning (Yang & Xu, 2018). Various definitions underscore the role of technology in smart cities. Giffinger et al. (2007) perceive smart cities as having the capacity to redistribute public resources, identify inadequate infrastructure, address municipal service allocations, and intervene in environmental issues. This perspective reflects a techno-utopian outlook on the positive changes smart cities could bring, although it remains elusive and desirable (Yang & Xu, 2018).

In a smart city driven by the IoT, interconnected devices, sensors, and technologies collaborate to enhance the efficiency and sustainability of urban living (Chavhan, 2022) The city strategically places a variety of sensors and devices, including smart streetlights, environmental sensors, surveillance cameras, and waste management sensors, to gather real-time data on aspects such as air quality, traffic flow, energy consumption, and waste levels (Chavhan, 2022).

This collected data traverses communication networks, such as wireless networks (Wi-Fi, cellular, Low Power Wide Area Network - LPWAN) or wired networks, to reach a central data processing system. Swift communication is essential for real-time decision-making and responsiveness in the smart city infrastructure (Wei et al., 2019). Subsequently, the data is transmitted to a centralized platform or decentralized (such as blockchain) or cloud-based infrastructure for processing and analysis (Ibrahim & Truby, 2022). Advanced analytics tools handle substantial data volumes, extracting meaningful insights and patterns to facilitate informed decisions for enhancing city operations (Wei et al., 2019).

Intelligent algorithms leverage the analyzed data to make decisions and trigger automated actions, such as adjusting traffic signal timings or optimizing energy consumption (Syed et al., 2021). Automation in processes like waste management, parking systems, and public transportation enhances efficiency and reduces resource consumption (Syed et al., 2021). Smart city initiatives often incorporate user interfaces, such as mobile apps or web portals, providing citizens with easy access to information and services (Kumar et al., 2020). Through these interfaces, citizens can provide feedback, report issues, and engage in community initiatives, fostering a more collaborative approach to city management (Kumar et al., 2020).

However, these network-connected sensors and actuators are often constrained by limited resources like processing power and memory (Firouzi et al., 2022). Then again, many scholars believe that with the rise of device capabilities, IoT can gather vast amounts of data from everyday technology use (Shafique et al., 2020). This data, originating from routine activities, blends with the digital realm, progressively becoming more personalized. The technological foundation of IoT remains neutral to the nature of collected and transferred data, lacking a clear distinction between personal and non-personal information (Chiara, 2021). For instance, identical network-connected sensor devices, like GPS, may handle data differently—one processing personal information (such as a person's location) and the other not (such as a parcel's location). Therefore, the context in which IoT devices operate and the intended purpose of collected data become pivotal elements in the privacy discourse (Brown et al., 2022).

Hence, individual privacy risks are significant, underscoring the importance of comprehending where data is captured, centralized, and processed, particularly when shared with governmental and corporate entities overseeing smart cities. Awareness of the potential impact of existing data collection and algorithmic profiling on societal regulation is increasing (Martucci et al., 2017), and it should be increasing. Current IoT applications illustrate how data collection is connected with corporate interests and people's daily activities. There are arguments about societal benefits, including safer roads, more effective medication regimes, and more sustainable energy use through IoT-embedded smart decision-making processes in smart cities.

However, these all add more value for corporations and less value for individuals and society. For example, digital medicines may link individuals' medication usage with pharmaceutical companies, and smart meters may connect household appliance usage with energy providers. By doing this, pharmaceutical companies can potentially monetize medication usage data by licensing it to research institutions, insurers, or other entities interested in health-related analytics. Similarly, data on household appliance usage can be valuable for market research, urban planning, or even selling anonymized data to third parties, providing an additional revenue stream for energy providers.

Nevertheless, it is always not that negative, and it does have value for society. By linking medication usage with pharmaceutical interests, smart cities can improve healthcare efficiency. Similarly, correlating household appliance usage with energy supplier interests allows for better resource management. The data collected from digital medicines and smart meters can give city planners valuable information to make informed decisions about infrastructure development, energy distribution, and healthcare facilities.

Hence, the application of IoT in urban settings is a two-way sword. Whatever it may be, there are important questions to answer about which norms and values are embedded in these systems, who has a say in how these applications are selected, and how an individual's data and right to privacy are protected.

Applications within the IoT can serve a variety of interests, but those with control over the infrastructure and data hold a significant advantage in prioritizing their concerns (Martucci et al., 2017). The *Thing Tank*, an IoT research project, has artfully critiqued the potential control dynamics within IoT through a satirical video (Vimeo, n.d). In this video, an elderly person navigates an IoT-centric life where their fork suggests meals, their bed dictates their routine, and their walking stick regulates exercise. This portrayal lays bare the impersonal control exerted by absent relatives through "assistive" IoT technologies, functioning as proxies for their concerns and desire for control. The video also reveals how these forms of non-consensual control incite rebellion, as the elderly person ingeniously circumvents each mechanism in subtle resistance. Moreover, it subtly conveys the value and joy derived from specific freedoms that overt device regulation seems to deny, encapsulating conflicting values within the elderly person, their relatives, and broader cultural narratives about responsible lifestyle choices and personal freedoms. Conspicuously absent in the centralized, technical, and bureaucratic IoT future depicted by the video is any consideration for creating space where these values can be negotiated (Martucci et al., 2017).

In addition to cultural narratives, political narratives play a vital role in shaping human perceptions of technology and their acceptance. "Sustainability" being another buzzword is a political narrative that easily can be sold for technology policy implementation. In the realm of IoT-enabled smart cities, digitally augmented transport networks, peer-to-peer energy trading, and other green projects promise environmental sustainability. Nevertheless, all these promises have privacy risks. Take the release of travel data for an example. Recently, we have seen media stories about Uber's inappropriate use of information. An Uber senior executive reportedly threatened to disclose aspects of journalists' private lives derived from Uber journey data in response to negative reporting (Martucci et al., 2017). Reports from alternative media have also illuminated instances in which Uber staff informally accessed, shared, and commented on passengers' journeys (Martucci et al., 2017). In response, Uber clarified that these practices were in contradiction to its privacy policies, and the executive who issued threatening remarks has issued an apology (Isaac, 2014). While Uber could not sustain the notion of being a playground for curious employees or a tool for a senior executive seeking retribution against critical journalists, these reports underscore the significant influence gained over users through the collection of journey data. If Uber achieves its goal of becoming the preferred option for all journeys, its extensive surveillance potential will be further heightened. This unequivocally highlights some of the privacy risks associated with IoT in smart city applications.

Beyond the direct implications for personal privacy, the applications of smart cities pose additional risks related to privacy concerns. These include the potential emergence of new "digital divides" (Karim, 2023), where opting out of non-privacy-friendly systems could lead to inequality of opportunity. There are also democratic risks, as aspects of public policy, such as the regulation of supply, demand, price, quality, and safety of the applications used in the smart city context, shift from local authorities to being (either directly or indirectly) regulated by the private corporate organizations.

Hence, the applications of smart cities have far-reaching effects on social principles, introducing potential challenges related to inequality, privacy, and democratic governance. Addressing these challenges requires careful consideration of the ethical and social implications of smart city technologies, ensuring that the benefits are distributed equitably and that democratic principles are upheld in the evolving urban landscape.

As such, policy discourse is a crucial component of this process. Developing and implementing effective policies is essential to guide the ethical use of technology, protect individual privacy, and establish a framework for responsible and transparent governance. Policies should address issues such as data collection, storage, sharing, and usage to strike a balance between the benefits of smart city technologies and the protection of citizens' rights. In the absence of well-defined policies and ethical considerations, there is a risk of potential misuse of data, infringement on individual privacy, and the concentration of power in the hands of a few entities. Therefore, in the next section of this chapter, I discuss how the policy discourse becomes a necessary step to navigate the complexities of smart city technologies while safeguarding the rights and interests of the citizens they are meant to serve.

The policy debate: Security and privacy

Smart cities pose various challenges, contrary to being a universal solution as discussed earlier in this chapter. Some of these challenges focus on technical aspects, such as capacity, technology access, data interoperability, and technical standardization, etc. (Sánchez-Corcuera et al., 2019). But, other than the technical challenges, there are also political challenges. Political obstacles involve the need to secure citizens' right to privacy and data protection. Despite proposing solutions like sensors, big data, and advanced computing, smart cities face criticism for prioritizing hardware over people (Edwards, 2016).

In the policy context, two pivotal issues emerge: security and privacy. Security pertains to the vulnerability of data to accidental or intentional breaches due to technical or organizational shortcomings. Privacy emphasizes individuals' rights to control the collection, processing, and subsequent reuse of their personal data.

The integration of smart city solutions into already intricate urban structures amplifies their complexity. Relying on wireless sensor networks and integrated communication systems makes cities susceptible to power failures, software errors, and cyber-attacks. The insecurity and vulnerability of smart city systems echo the well-known lack of security and trustworthiness in the IoT (Maple, 2017). The inherent insecurity of the IoT arises from its small size, low cost, lack of independent power, and historical design for industrial rather than consumer use (Edwards, 2016). Common issues in IoT devices include poor encryption strength and a lack of security features (Aqeel et al., 2022). The reliance on wireless communications without mandatory technical and security standards makes them susceptible to attacks. For smart cities, these security challenges are compounded by the complexities of multiple vendors and interoperating systems. Many cities implement new technologies with minimal cybersecurity testing, leaving them vulnerable to attacks (Edwards, 2016).[1] Unlike conventional software, vulnerabilities in embedded architectures cannot be easily solved, posing a potential future threat known as the "Internet of Junk" (Edwards, 2016).

It remains a question whether privacy laws or policy implementation relating to data protection can effectively address the imminent security challenges of smart cities, as discussed above. Many data protection laws around the world provide a specific proactive measure that addresses security concerns through the mandate for security breach disclosure. For example, in Europe, this obligation is ensured through the General Data Protection Regulation (GDPR). In cases of data breaches reaching a certain severity, privacy regulators must be notified. Yet, under many data protection laws, data controllers[2] may assert a defense if they can demonstrate the implementation of adequate security measures.

Additionally, privacy laws can mandate the principle of data minimization, which ensures that only necessary data is collected and retained. Similarly, purpose limitation clauses can restrict the use of data to specific, predefined purposes, preventing unauthorized access and misuse. Most of the privacy laws around the world enforce stringent consent requirements, ensuring that individuals are fully informed about the data being collected, its purpose, and who has access to it. Modern privacy laws also require transparency in data processing activities, making it mandatory for organizations to disclose their data practices to the public. Privacy laws also grant individuals specific rights over their data, such as the right to access, correct, or delete personal information.

However, the question remains whether all these recent privacy law developments or the legal principles of data protection are sufficient for the data security of smart cities or not. Eventually, many scholars believe it is not sufficient at all (Choenni et al., 2022; Calzada, 2022). Modern privacy laws should specify security standards and measures that smart city systems must adhere to, including encryption, access controls, and regular security audits. A notable challenge surfaces in the absence of global harmonization in legal standards for security, privacy, and data protection. As smart city technologies often involve cross-border data flows, privacy and data protection laws should include provisions for international data protection standards and cooperation among jurisdictions to address security concerns.

While in Europe, the Budapest Cybercrime Convention offers a foundational level of international harmonization for cybersecurity, its primary focus is on facilitating global law enforcement in criminal matters rather than elevating security standards for the industry. Importantly, it does not mandate civil liability, though Article 13 allows for its existence. Although beyond the scope of this article, exploring the extension of provisions safeguarding critical infrastructure to smart cities introduces an intriguing avenue for future investigation.

Then again, within the European Union (EU), the regulation of IoT security has been marked by the prevalence of "soft law" over hard law (Edwards, 2016). A notable instance is the establishment of a specialized yet non-mandatory Privacy Impact Assessment (PIA) procedure for radio frequency identification (RFID) chip installations, essentially an early subset of the IoT. This development aligns with the broader regulatory trend encouraging a proactive approach to security risks with privacy by design (PbD) principles (Edwards, 2016).

In addition to the PbD, privacy laws, in the context of smart cities, should incorporate ethical considerations, ensuring that the deployment of smart city technologies aligns with societal values and norms. Privacy laws can also mandate public awareness programs to educate citizens about the risks and benefits of smart city technologies. Informed citizens are better equipped to protect their privacy and demand accountability from public and private entities.

These policy matters demonstrate that the security of smart cities closely depends on how data protection laws are enforced. As the enactment of laws, security procedures, and willingness to engage public perception depend on the policy discourse of a specific government, politics plays a crucial role in how a government wants to utilize smart city applications on its citizens. As mentioned earlier, it is a two-sided sword, not only for individual citizens but also for the government. The government can enhance surveillance over citizens through these smart city applications, or it can provide a better lifestyle by keeping privacy and data protection intact with proper regulatory steps.

The decision rests on the willingness of the government and the system that needs to be implemented. While such implementation requires a lot of technical and legal discourse, it is also a policy matter that a democratic society should debate upon.

Legal and regulatory aspects of individuals' data collection in the context of smart cities

In this section, I explore the legal aspects related to the data protection regulation, with a specific focus on GDPR, concerning the handling of personal information in the realms of IoT and smart cities. Data protection laws specifically apply to information categorized as personal data. According to the GDPR, personal data is defined as any information related to an identified or identifiable natural person (data subject), who can be recognized directly or indirectly through identifiers such as a name, identification number, location data, online identifier, or factors specific to their physical, physiological, genetic, mental, economic, cultural, or social identity.

The GDPR definition emphasizes that unique device identifiers, such as media access control (MAC) addresses or RFID tag codes, may also be considered as personal data linked to users. This implies that users could potentially be profiled under these device identifiers by observers, even without direct identification by name. For example, an RFID tag in a regularly used smartwatch might be used in a supermarket to create a user profile as a returning customer. The European Court of Justice (ECJ) has asserted that Internet Protocol (IP) addresses merit protection as personally identifiable information because they enable the precise identification of users (Mahieu et al., 2019). Similar considerations are being contemplated for MAC addresses. Even in the United States, the Children's Online Privacy Protection Act (COPPA) unequivocally designates IP addresses, MAC addresses, and other unique device identifiers as personal information (Moura Vicente & de Vasconcelos Casimiro, 2021). The Federal Trade Commission (FTC) revised the COPPA Rule in 2013 to explicitly include these identifiers within the definition of "personal information". The Health Insurance Portability and Accessibility Act (HIPAA) also considers device identifiers, such as MAC addresses, IP addresses, and RFID tags, as "identifiers" requiring removal for the de-identification of protected health information.

In the context of IoT and smart sensing, individuals can often be identified through data from devices that they use (smartwatches, automatic cars, smart house systems, smart appliances, etc.), revealing lifestyle aspects of individuals and families. Sensors of these devices are designed to collect data without explicit, case-by-case consent, introducing transparency challenges. Furthermore, the amalgamation of sensor data can unveil sensitive personal details, such as habits and physical condition. The collected data and derived information can also vary widely, each serving different purposes. This complexity makes obtaining informed consent and setting precise privacy preferences for various data types and purposes a demanding task for end users.

In terms of the legal aspects of smart cities in Europe, at least 15 articles of the GDPR are worth further discussion. Article 5 of the GDPR lays out key principles for collecting personal data, emphasizing the need for clear purposes. In the context of smart cities, this means that personal data should be gathered for specific and lawful reasons, and any subsequent use must align with these predefined goals (Löfgren & Webster, 2020). This helps prevent mishandling or unauthorized use of personal information, ultimately enhancing overall privacy protection. Moreover, Article 5 of the GDPR underscores the importance of data minimization, especially crucial in the context of smart cities where a significant amount of data is generated. It is essential to carefully consider how much data is collected and processed. The GDPR promotes the idea of minimizing the collection and processing of personal data and encourages the use of anonymization or pseudonymization when possible. This ensures that only necessary and relevant data is collected, reducing potential risks associated with excessive data processing.

Additionally, Article 25 of the GDPR mandates data controllers to implement suitable technical and organizational measures to ensure privacy by default. Essentially, this means that smart city systems should be designed and set up from the outset to adhere to privacy principles. This includes measures that uphold purpose specification, data minimization, and other crucial privacy requirements. All these measures work together to achieve the overarching goal of protecting individuals' privacy rights in the smart city environment.

Furthermore, the GDPR, as outlined in Articles 6 and 7, underscores the critical importance of ensuring data processing with informed consent. In the dynamic landscape of smart cities, where a substantial amount of personal data is collected, obtaining explicit consent is crucial, especially when dealing with sensitive information such as health, ethnicity, or political opinions (Löfgren & Webster, 2020). This necessitates smart city developers and administrators to establish transparent and distinct methods for obtaining and managing consent.

In addition, Article 15 grants data subjects the right to access their personal data and obtain information about processing purposes, data recipients, retention periods, and more. Articles 16, 17, and 18 empower data subjects to correct or erase their data and limit its processing. In the intricate environment of smart cities, where data is continuously generated and processed, it is essential to establish efficient mechanisms for citizens to exercise these rights (Hashem et al., 2016; Silva et al., 2018). This involves creating automated systems that enable quick responses to correction and deletion requests.

Articles 21 and 22 of the GDPR confer upon individuals the right to challenge data processing, particularly in scenarios involving automated decision-making, including profiling. In the context of smart cities, where automated systems shape various aspects of urban life, citizens should possess the ability to contest decisions made through automated processes. This not only ensures accountability but also prevents potential biases in algorithmic decision-making.

The GDPR, along with numerous contemporary global data privacy laws, plays a pivotal role in shaping privacy and data protection within the continually evolving realm of smart cities. Adhering to the diverse articles outlined in the GDPR empowers smart city administrators to cultivate an environment that upholds individual rights, ensures transparency, and establishes robust security measures. Striking a balance between the advantages of smart city technologies and the safeguarding of citizens' privacy demands a meticulous and comprehensive implementation of data protection principles and regulations.

While the legislation delineates legal definitions, privacy requirements, and consequences for violations, it lacks the technological capacity to enforce data protection. Relying solely on legal measures for privacy in smart cities proves insufficient (Christofi et al., 2019). Therefore, essential support from computer and network security tools, collectively known as privacy-enhancing technologies (PETs), is imperative (Safa et al., 2022). PETs are categorized into three groups based on their goals (Martucci et al., 2017). The first category focuses on enforcing the legal privacy principle of data minimization, incorporating elements like anonymity, unlinkability, pseudonymity, and unobservability (Martucci et al., 2017). Privacy by design (PbD), aligned with GDPR Art. 23, underscores the default implementation of these PETs, applicable at the network or application levels. PbD is a concept that promotes the integration of privacy and data protection principles into the design and development of systems, processes, and technologies, rather than addressing privacy as an afterthought or add-on. The goal is to embed privacy features and considerations into the initial stages of the design process, making privacy an integral part of the overall system.

Conclusion

The applications of smart cities have far-reaching effects on social principles, introducing potential challenges related to inequality, privacy, and democratic governance. Addressing these challenges requires careful consideration of the ethical and social implications of smart city technologies, ensuring that the benefits are distributed equitably and that democratic principles are upheld in the evolving urban landscape. Additionally, the policy debate surrounding security and privacy in smart cities highlights the multifaceted challenges faced by these urban ecosystems. While technical challenges such as capacity, data interoperability, and standardization are significant, political obstacles related to citizens, right to privacy, and data protection also come to the forefront.

While modern privacy laws in different jurisdictions provide a foundation for protecting individuals' data, the current legal developments may not be sufficient to address smart city security challenges. Modern privacy laws should specify security standards, encryption, access controls, and regular security audits tailored to smart city systems. The absence of global standards necessitates international cooperation and provisions in privacy laws for addressing security concerns.

As such, the legal aspects of data protection play a crucial role in shaping privacy within smart cities. The principles of purpose specification, data minimization, and consent from data subject, by default, provide a framework for responsible data handling. However, the legislation alone is insufficient, necessitating the incorporation of PETs to enforce these principles. Additionally, public awareness programs, ethical considerations, and transparency initiatives should be incorporated into privacy laws to empower citizens and maintain a balance between technological advancement and individual privacy. The implementation of these recommendations requires a collaborative effort between governments, technologists, and citizens in an ongoing policy discourse.

Notes

1 Traffic control sensors installed in Washington DC, New York, London, Lyon, and other cities can be easily attacked with a simple exploit programmed on cheap hardware.

2 A data controller is a "person, public authority, agency or any other body which… determines the purposes and means of the processing of personal data".

References

Aqeel, M., Ali, F., Iqbal, M. W., Rana, T. A., Arif, M., & Auwul, M. R. (2022). A review of security and privacy concerns in the internet of things (IoT). *Journal of Sensors*, *2022*, 1–20.

Brauneis, R., & Goodman, E. P. (2018). Algorithmic transparency for the smart city. *Yale Journal of Law & Technology*, *20*, 103.

Brown, R., Truby, J., & Ibrahim, I. A. (2022). Mending Lacunas in the EU's GDPR and proposed artificial intelligence regulation. *European Studies*, *9*(1), 61–90.

Calzada, I. (2022). Citizens' data privacy in China: The state of the art of the Personal Information Protection Law (PIPL). *Smart Cities*, *5*(3), 1129–1150.

Chavhan, S. (2022). Shift to 6G: Exploration on trends, vision, requirements, technologies, research, and standardization efforts. *Sustainable Energy Technologies and Assessments*, *54*, 102666.

Chiara, P. G. (2021). The balance between security, privacy and data protection in IoT data sharing: A critique to traditional "Security&Privacy" surveys. *European Data Protection Law Review*, 7, 18.

Choenni, S., Bargh, M. S., Busker, T., & Netten, N. (2022). Data governance in smart cities: Challenges and solution directions. *Journal of Smart Cities and Society*, *1*(1), 31–51.

Christofi, A., Heyman, R., Vandercruysse, L., Verdoodt, V., Buts, C., Dooms, M., Pierson, J. & Valcke, P. (2019). Smart city privacy: Enhancing collaborative transparency in the regulatory ecosystem. In *2019 CTTE-FITCE: Smart cities & information and communication technology (CTTE-FITCE)* (pp. 1–5). IEEE.

de Gracia-Soriano, P., Jareño-Ruiz, D., Jiménez-Delgado, M., & Jiménez-Delgado, A. (2021). Sociological reflections on smart cities and city times. In *Smart cities and the UN SDGs* (pp. 197–209). Elsevier.

Edwards, L. (2016). Privacy, security and data protection in smart cities: A critical EU law perspective. *European Data Protection Law Review*, *2*, 28.

Firouzi, F., Farahani, B., & Marinšek, A. (2022). The convergence and interplay of edge, fog, and cloud in the AI-driven Internet of Things (IoT). *Information Systems*, *107*, 101840.

Giffinger, R., Fertner, C., Kramar, H., Kalasek, R., Pichler-Milanovie, N., & Meijers, E. (2007). Smart cities: Ranking of European medium-sized cities, Vienna, Austria: Centre of REGIONAL science (SRF), VIENNA University of Technology, https://www.smart-cities.eu/download/smart_cities_final_report.pdf

Hashem, I. A. T., Chang, V., Anuar, N. B., Adewole, K., Yaqoob, I., Gani, A., Ahmed, E. & Chiroma, H. (2016). The role of big data in smart city. *International Journal of Information Management*, *36*(5), 748–758.

Hu, M., & Li, C. (2012). Design smart city based on 3S, Internet of things, grid computing and cloud computing technology. *IOT Workshop, CCIS*, *312*, 466–472.

Huang, H., Yao, X. A., Krisp, J. M., & Jiang, B. (2021). Analytics of location-based big data for smart cities: Opportunities, challenges, and future directions. *Computers, Environment and Urban Systems*, *90*, 101712.

Ibrahim, I. A., & Truby, J. (2022). Governance in the era of Blockchain technology in Qatar: A roadmap and a manual for trade finance. *Journal of Banking Regulation*, *23*(4), 419–438.

Isaac, M. (2014). "Uber executives comments leave company scrambling". Bits, a NY Times Blog. https://archive.nytimes.com/bits.blogs.nytimes.com/2014/11/18/emil-michael-of-uber-proposes-digging-into-journalists-private-lives/

Ismagilova, E., Hughes, L., Rana, N. P., & Dwivedi, Y. K. (2020). Security, privacy and risks within smart cities: Literature review and development of a smart city interaction framework. *Information Systems Frontiers*, *24*, 393–414.

Karim, R. (2023). Toward an International Law of Just AI Development. *Journal of East Asia and International Law*, *16*(2), 251–266.

Kitchin, R. (2016). The ethics of smart cities and urban science. *Philosophical Transactions of the Royal Society A: Mathematical, Physical and Engineering Sciences*, *374*(2083), 20160115.

Kumar, H., Singh, M. K., Gupta, M. P., & Madaan, J. (2020). Moving towards smart cities: Solutions that lead to the Smart City Transformation Framework. *Technological Forecasting and Social Change*, *153*, 119281.

Lai, C. S., Jia, Y., Dong, Z., Wang, D., Tao, Y., Lai, Q. H., Wong, R.T.K., Zobaa, A.F., Wu, R. & Lai, L. L. (2020). A review of technical standards for smart cities. *Clean Technologies*, *2*(3), 290–310.

Löfgren, K., & Webster, C. W. R. (2020). The value of Big Data in government: The case of 'smart cities'. *Big Data & Society*, *7*(1), 2053951720912775.

Mahieu, R., Van Hoboken, J., & Asghari, H. (2019). Responsibility for data protection in a networked world: On the question of the controller, effective and complete protection and its application to data access rights in Europe. *Journal of Intellectual Property, Information Technology and E-Commerce Law*, *10*, 84.

Maple, C. (2017). Security and privacy in the internet of things. *Journal of Cyber Policy*, *2*(2), 155–184.

Martucci, L. A., Fischer-Hübner, S., Hartswood, M., & Jirotka, M. (2017). Privacy and social values in smart cities. In *Designing, Developing, and Facilitating Smart Cities: Urban Design to IoT Solutions*, edited by V. Angelakis, E. Tragos, H. Pöhls, A. Kapovits, and A. Bassi, 89–107.

Miorandi, D., Sicari, S., De Pellegrini, F., & Chlamtac, I. (2012). Internet of things: Vision, applications and research challenges. *Ad hoc Networks*, *10*(7), 1497–1516.

Moura Vicente, D., & de Vasconcelos Casimiro, S. (2021). Data protection in the Internet. In *General reports of the XXth general congress of the international academy of comparative law-Rapports généraux du XXème Congrès général de l'Académie internationale de droit comparé* (pp. 611–634). Springer International Publishing.

Nam, T., & Pardo, T. A. (2011). Conceptualising smart city with dimensions of technology, people, and institutions. *The Proceedings of the 12th Annual International Conference on Digital Government Research*, June 12–15, 282–291.

Safa, N. S., Mitchell, F., Maple, C., Azad, M. A., & Dabbagh, M. (2022). Privacy Enhancing Technologies (PETs) for connected vehicles in smart cities. *Transactions on Emerging Telecommunications Technologies*, *33*(10), e4173.

Sánchez-Corcuera, R., Nuñez-Marcos, A., Sesma-Solance, J., Bilbao-Jayo, A., Mulero, R., Zulaika, U., Azkune, G. & Almeida, A. (2019). Smart cities survey: Technologies, application domains and challenges for the cities of the future. *International Journal of Distributed Sensor Networks*, *15*(6), 1550147719853984.

Shafique, K., Khawaja, B. A., Sabir, F., Qazi, S., & Mustaqim, M. (2020). Internet of things (IoT) for next-generation smart systems: A review of current challenges, future trends and prospects for emerging 5G-IoT scenarios. *IEEE Access*, *8*, 23022–23040.

Silva, B. N., Khan, M., & Han, K. (2018). Towards sustainable smart cities: A review of trends, architectures, components, and open challenges in smart cities. *Sustainable Cities and Society*, *38*, 697–713.

Sookhak, M., Tang, H., He, Y., & Yu, F. R. (2018). Security and privacy of smart cities: A survey, research issues and challenges. *IEEE Communications Surveys & Tutorials*, *21*(2), 1718–1743.

Syed, A. S., Sierra-Sosa, D., Kumar, A., & Elmaghraby, A. (2021). IoT in smart cities: A survey of technologies, practices and challenges. *Smart Cities*, *4*(2), 429–475.

Vimeo (n.d.). The "Uninvited Guests" is short film produced by Superflux and commissioned by Thing Tank. https://vimeo.com/128873380

Wei, X., Duan, Q., & Zhou, L. (2019). A QoE-driven tactile Internet architecture for smart city. *IEEE Network*, *34*(1), 130–136.

Yahia, N. B., Eljaoued, W., Saoud, N. B. B., & Colomo-Palacios, R. (2021). Towards sustainable collaborative networks for smart cities co-governance. *International Journal of Information Management*, *56*, 102037.
Yang, F., & Xu, J. (2018). Privacy concerns in China's smart city campaign: The deficit of China's Cybersecurity Law. *Asia & the Pacific Policy Studies*, *5*(3), 533–543.
Yu, W., & Xu, C. (2018). Developing smart cities in China: An empirical analysis. *International Journal of Public Administration in the Digital Age*, *5*(3), 76–91.

CONCLUSION

Esmat Zaidan, Imad Antoine Ibrahim, and Elie Azar

The world is undergoing a technology-driven transformation aimed at enhancing the preservation, revitalization, livability, and sustainability of urban areas. This shift is reflected in various government initiatives and efforts to promote and adopt smart, efficient solutions, along with significant investments in Information and Communication Technology (ICT) infrastructure and advanced technologies to foster the development of smart cities (Chatti & Majeed, 2022; Rao, 2007). These cities facilitate tangible cooperation between governments and citizens through diverse initiatives to ensure sustainable urban management. As more people migrate to cities in search of better opportunities, the pressure on existing resources increases (Chang et al., 2020). In theory, these new types of cities rely on data-driven decision-making, which holds the promise of creating safer communities, reducing environmental footprints, and boosting efficiency across key sectors through digitalization (Serrano, 2018; De Dutta & Prasad, 2020).

This effort is part of a highly publicized initiative to reshape urban development through megaprojects, centralized master planning, and strategies that project modernity and global prosperity. However, transitioning a city into a fully smart urban environment is a complex and large-scale undertaking, with challenges expected to multiply significantly (Yigitcanlar, 2015; Fadhel et al., 2024). Expanding this transition across an entire city also introduces social and policy-related issues on a broader scale. In this context, a solution to achieving a citywide or regional smart transformation may lie in incremental planning and phased prioritization of smart urban communities to meet high-level objectives and eventually form an interconnected system (Bibri, 2021; Lee et al., 2013). A comprehensive approach is crucial due to the existing challenges. The focus should be on creating an ecosystem that

DOI: 10.4324/9781003439325-16

considers all relevant factors, much like a natural ecosystem. Otherwise, this new attempt to reorganize cities within the framework of the Fourth Industrial Revolution is likely to fail (Silva de Amarim et al., 2019; Yun & Lee, 2019).

Like globalization and industrialization, the concept of smart cities is poised to bring significant global transformations, though each nation will experience these changes differently based on its unique conditions, ecosystems, resources, and challenges (Hayat, 2016). Terms such as "intelligent" and "knowledge cities" also focus on sustainability and innovation (Anttiroiko et al., 2014). The idea of moving from a smart city to a smart society is not new; it involves redefining what "smart" means, adopting new rules, and addressing societal needs alongside technological innovation (Valkenburg et al., 2016). The expansion of this concept stems from the need to address sustainability in its social, economic, and environmental dimensions within a digital and knowledge-driven world.

This approach spans sectors like business, healthcare, and transportation (Mehmood et al., 2022) aiming to create a sustainable society with improved welfare, health, and quality of life (Chen, 2022). Central to this vision is the implementation of smart systems, such as smart energy and technology (Cifaldi & Serban, 2018), to manage resources efficiently while enhancing quality of life. However, it remains uncertain whether this objective is achievable (Heidari et al., 2022), as many factors influence the outcome. For example, recent initiatives by the Japanese government (Deguchi, 2020), focused on providing goods and services to its citizens, have revealed gaps in reaching the so-called "Super Smart Society" (Iqbal & Olariu, 2021).

The relationship between smart cities and technology is crucial, particularly considering the role of innovation in addressing the urgency of the mission (Roche et al., 2012). Yet not everyone supports the concept. Critics argue that smart cities are incompatible with the informal nature of urban life, risk exacerbating social and urban inequalities, and often empower corporations at the expense of citizens (Krivy, 2016). Moreover, some claim that smart cities do not adequately leverage human capital, reducing their effectiveness (Oliveira & Campolargo, 2015), as rapid technological advances outpace their potential (Perez-DelHoyo & Mora, 2019).

Some argue that the success of smart cities depends on various factors, including advanced ecological and technological systems, existing socio-cultural structures, and scientific discourse (Bibri & Krogstie, 2017). This book addresses these complexities by providing perspectives from the Global North and South, focusing on citizens, governance, and the law. The contributors highlight challenges but also offer solutions, adding value to the broader goal of creating both smart cities and smart societies while ensuring a smooth transition. The editors, however, acknowledge that their focus on three specific aspects limits the scope of the book, as many other important issues remain unaddressed.

Citizens hold varying opinions with some advocating for further progress and with others expressing dissatisfaction, both passively and actively. In the case of the former, citizens often lack the awareness and skills to effectively voice their opinions. In the latter, individuals are discontent with the current state of technology, the political system, and the overall societal impact. As a response, governments are working to shift the narrative by promoting the right technologies, laws, and information to engage citizens (Bibri & Krogstie, 2017). Changing public perception is crucial, as, by the end of the century, most people will be living in cities, necessitating better collaboration between the public and private sectors (Georgiadis et al., 2021). Frameworks are being developed to assess citizen participation across various cities and societies globally (Simonofski et al., 2017).

This is particularly important, as not every citizen can be considered "smart" in the context of urban development, despite their essential role (Ferronato & Ruecker, 2018). Citizens themselves are evolving alongside the urban landscape (Vanolo, 2016), with the ultimate goal of making cities smarter through meaningful input and added value to urban development (Castelnovo, 2016). However, increasing skepticism toward new technologies and their societal role remains, despite efforts to highlight their benefits (Bull & Azennoud, 2016). Valid concerns exist, such as fears of technology infringing on citizens' freedoms and well-being (Vanolo, 2016). Additionally, the role of citizens is often more complex and ambiguous than it appears in practice (Shetlon & Lodato, 2019).

Stokseth, Borgebund, and Sætra effectively highlight these realities and provide recommendations for enhancing citizen participation in various aspects. A holistic approach is required, considering different fields, as emphasized by Smit et al. in their focus on digital literacies. In recent years, fields like artificial intelligence have gained relevance, with Khutsishivili focusing on its role in empowering citizens. Meanwhile, Lee explored top-down and bottom-up approaches to involving citizens.

The governance aspect plays an equally important role in establishing a smart future, which is why many authors have explored the relationship between smart cities, communities, and governance. Their focus has been on addressing governance in the context of emerging technologies such as big data and the Internet of Things, along with the role of people, institutions, corporations, and e-governance in shaping the future of cities and their impact on the environment (Kuzior et al., 2023). These studies have led to calls for new, more collaborative governance models for smart cities, aimed at overcoming current challenges and capitalizing on opportunities as investments are made (Barns, 2018). Such calls attempt to address fragmented approaches as academic interest in governance increases, especially from sociotechnical perspectives (Meijier & Boliva, 2015).

The goal is to achieve smart governance that integrates environmental sustainability, mobility, living standards, people, and the economy through public participation, effective services, and transparency (Anand & Navia-Marco, 2018). While technology can facilitate collaboration between citizens and local governments to advance these goals, there is currently limited empirical evidence to support this potential (Tomor et al., 2019). Therefore, risk management is viewed as essential for a sustainable smart city (Ullah et al., 2021), requiring adaptive governance practices, effective arrangements (Maccani et al., 2020), and collaboration among diverse actors through specific instruments and technologies (Ooms et al., 2020). Nevertheless, it remains unclear how governance can be fully integrated into the framework of a smart city, given the various forms it can take (Allam & Peter Newman, 2018).

Antunes et al. addressed the governance issue, proposing a multifaceted solution for the use of public data. Meanwhile, Reshetova and Hartley discussed ways to enhance accountability calling for new scholarly conversations on governance reforms. Ingallina, Park, and Carlotti tackled eco-technological innovation in Morocco and South Korea with proper recommendations for urban projects. Finally, Abdouni highlighted the ways bottom-up interventions informally can smartify cities.

The final element discussed in the book is the role of law in achieving smart cities and societies. Legislation reflects the values of its time, follows specific processes for implementation, and focuses on adopting and enforcing rules with the goal of realizing the rule of law (Keymolen & Voorwinden, 2020). However, many caution that the law may support market mechanisms at the expense of addressing social inequality (Kummitha & Nathalie Crutzen, 2017) and protecting human rights, which may be threatened by emerging technologies (Brown, 2019). It is worth noting that the legal discipline developed relatively late compared to other fields, as seen in regulatory frameworks like cybersecurity law, artificial intelligence, and energy regulations (He et al., 2022).

New forms of rules may be needed to safeguard society, extending beyond traditional legal approaches, such as design solutions or technologies for public control, given the law's limitations (Gasiola et al., 2019). Literature to consider includes interdisciplinary studies and reports on urban law, behavioral regulation, and economics (Ranchordas, 2020). The legal field faces the challenge of requiring lawmakers to fully understand these developments before drafting laws for smart cities or societies. This is particularly difficult since lawyers often lack technical expertise, and the legislative process can take years. Further research is needed on the role of regulation to ensure good governance of smart cities and societies, although some progress has been made in the last decade (Jewell, 2018).

Jiang and Zheng tackled specific aspects in relation to this topic focusing on the geopolitics of international law. Meanwhile, Hennaoui focused on sustainable urbanization from a regulatory perspective. Karim addressed privacy and data collection, while Adebayo and Ibrahim studied regulatory compliance in coastal cities in Nigeria.

Overall, the added value of the research lies in its focus on a comparative approach of developed and developing nations simultaneously as the number of such cities in the South is constantly increasing. A smart city must reflect the circumstances and situations and consider the population and any factor of relevance in the place where it is being created. In that sense, there is no magic formula or one-size-fits-all model that can be applied everywhere. This goes against the concept itself and against years of research highlighting the negative consequences of simply copying policies from the West and applying them to developing countries. New frameworks may need to be developed for smart city implementation in the Global South, drawing on recent data and research that can also benefit developed nations.

A comparative analysis of smart cities is already underway, examining the design, governance, and implementation of the concept in cities such as Amsterdam, Barcelona, Dubai, and Abu Dhabi. Through these studies, stakeholders are identified, governance models evaluated, and valuable lessons drawn (Noori et al., 2020). Similar analyses have also been conducted for several Chinese cities, highlighting the diverse approaches taken to achieve the same goal (Ma, 2021). These studies provide a foundation for developing holistic legal frameworks that consider the unique contexts and specific characteristics of each city, while acknowledging that the smart city concept itself is continually evolving (Gracias et al., 2023; Zhao et al., 2021).

References

Allam, A., & Peter Newman, P. (2018). Redefining the smart city: Culture, metabolism and governance. *Smart Cities*, *1*(1), 4–25.

Anand, P. B., & Navia-Marco, J. (2018). Governance and economics of smart cities: Opportunities and challenges. *Telecommunications Policy*, *42*(10), 795–799.

Anttiroiko, A. V., Valkama, P., & Bailey, S. J. (2014). Smart cities in the new service economy: Building platforms for smart services. *AI & Society*, *29*, 323–334.

Barns, S. (2018). Smart cities and urban data platforms: Designing interfaces for smart governance. *City, Culture and Society*, *12*, 5–12.

Bibri, S. E., & Krogstie, J. (2017). On the social shaping dimensions of smart sustainable cities: A study in science, technology, and society. *Sustainable Cities and Society*, *29*, 219–246.

Bibri, S. E. (2021). Data-driven smart sustainable cities of the future: Urban computing and intelligence for strategic, short-term, and joined-up planning. *Computational Urban Science*, *1*(8), 1–29.

Brown, T. E. (2019). Human rights in the smart city: Regulating emerging technologies in city places. In *Regulating new technologies in uncertain times* (pp. 47–65). Springer.

Bull, R., & Azennoud, M. (2016). Smart citizens for smart cities: Participating in the future. *Proceedings of the Institution of Civil Engineers – Energy*, *169*(3), 1–18.

Castelnovo, W. (2016). Co-production makes cities smarter: Citizens participation in smart city initiatives. In *Co-production in the Public sector: Experiences and challenges* (pp. 97–117). Springer.

Chang, V., Sharma, S., & Li, C. S. (2020). Smart cities in the 21st century. *Technological Forecasting and Social Change*, *153*, 119447.

Chatti, W., & Majeed, M. T. (2022). Information communication technology (ICT), smart urbanization, and environmental quality: Evidence from a panel of developing and developed economies. *Journal of Cleaner Production*, *366*, 132925.

Chen, C. W. (2022). From smart cities to a happy and sustainable society: Urban happiness as a critical pathway toward sustainability transitions. *Local Environment*, *27*(12) 1536–1540.

Cifaldi, G., & Serban, I. (2018). Smart cities-smart societies. In *Intelligent human systems integration* (pp. 700–707). Springer.

De Dutta, S., & Prasad, R. (2020). Digitalization of global cities and the smart grid. *Wireless Personal Communications*, *113*, 1385–1395.

Deguchi, A. (2020). *From smart city to society 5.0, in society 5.0: A people-centric super-smart society*. Springer.

Fadhel, M. A. et al. (2024). Comprehensive systematic review of information fusion methods in smart cities and urban environments. *Information Fusion*, *107*, 1–34.

Ferronato, P., & Ruecker, S. (2018). Citizenship: Designing the interaction between citizens and smart cities. In *Design as a catalyst for change - DRS International Conference 2018*, 25–28 June, Limerick, Ireland.

Gasiola, G. G. et al. (2019). Smart cities through smart regulation [Opinion]. *IEEE Technology and Society Magazine*, *38*(1), 25–28.

Georgiadis, A., Christodoulou, P., & Zinonos, Z. (2021). Citizens perception of smart cities: A case study. *Applied Sciences*, *11*(6), 1–20.

Gracias, J. S. et al. (2023). Smart cities—A structured literature review. *Smart Cities*, *6*(4), 1719–1743.

Hayat, P. (2016). Smart cities: A global perspective. *India Quarterly: A Journal of International Affairs*, *72*(2), 177–191.

He, W., Li, W., & Deng, P. (2022). Legal governance in the smart cities of China: Functions, problems, and solutions. *Sustainability*, *14*(15), 1–24.

Heidari, A., Navimipour, N. J., & Unal, M. (2022). Applications of ML/DL in the management of smart cities and societies based on new trends in information technologies: A systematic literature review. *Sustainable Cities and Society*, *85*, 104089.

Iqbal, A., & Olariu, S. (2021). A survey of enabling technologies for smart communities. *Smart Cities*, *4*(1), 54–77.

Jewell, M. (2018). Contesting the decision: Living in (and living with) the smart city. *International Review of Law, Computers & Technology*, *32*(2/3), 210–229.

Keymolen, E., & Voorwinden, A. (2020). Can we negotiate? Trust and the rule of law in the smart city paradigm. *International Review of Law, Computers & Technology*, *34*(3), 233–253.

Krivy, M. (2016). Towards a critique of cybernetic urbanism: The smart city and the society of control. *Planning Theory*, *17*(1), 8–30.

Kummitha, R. K. R., & Nathalie Crutzen, N. (2017). How do we understand smart cities? An evolutionary perspective. *Cities*, *67*, 43–52.

Kuzior, A. et al. (2023). E-Governance in smart cities: Global trends and key enablers. *Smart Cities*, *6*(4) 1663–1689.

Lee, J. H., Phaal, R., & Lee, S. H. (2013). An integrated service-device-technology roadmap for smart city development. *Technological Forecasting and Social change*, *80*(2), 286–306.

Ma, W. (2021). From city promotion via city marketing to city branding: Examining urban strategies in 23 Chinese cities. *Cities*, *116*, 103269.

Maccani, G. et al. (2020). An emerging typology of IT governance structural mechanisms in smart cities. *Government Information Quarterly*, *37*(4), 101499.

Meijier, A., & Boliva, M. P. R. (2015). Governing the smart city: A review of the literature on smart urban governance. *International Review of Administrative Sciences*, *82*(2), 392–408.

Mehmood R. et al. (2022). Editorial: Smart societies, infrastructure, systems, technologies, and applications. *Mobile Networks and Applications*, *28*, 598–602.

Noori, N., Hoppe, T., de Jong, M. (2020). Classifying pathways for smart city development: Comparing design, governance and implementation in Amsterdam, Barcelona, Dubai, and Abu Dhabi. *Sustainability*, *12*(10), 4030.

Oliveira, A., & Campolargo, M. (2015). From smart cities to human smart cities. In *2015 48th Hawaii international conference on system sciences, Kauai, HI, USA, 2015* (pp. 2336–2344). Hawaii.

Ooms, W. et al. (2020). Ecosystems for smart cities: Tracing the evolution of governance structures in a dutch smart city initiative. *International Entrepreneurship and Management Journal*, *16*, 1225–1258.

Perez-DelHoyo, R., & Mora, H. (2019). Knowledge society technologies for smart cities development. In *Smart cities: Issues and challenges* (pp. 185–198) Elsevier.

Rao, N. H. (2007). A framework for implementing information and communication technologies in agricultural development in India. *Technological Forecasting and Social Change*, *74*(4), 491–518.

Ranchordas, S. (2020). Nudging citizens through technology in smart cities. *International Review of Law, Computers & Technology*, *34*(3), 254–276.

Roche, S. et al. (2012). Are 'Smart Cities' smart enough. In A. Rajabifard & D Coleman (Eds.), *Spatially enabling government, industry and citizens: Research and development perspectives*. (pp. 215–235). GSDI Association Press

Serrano, W. (2018). Digital systems in smart city and infrastructure: Digital as a service. *Smart Cities*, *1*(1), 134–154.

Shetlon, T., & Lodato, T. (2019). Actually existing smart citizens. *City*, *23*, 35–52.

Silva de Amarim et al. (2019). Urban challenges and opportunities to promote sustainable food security through smart cities and the 4th industrial revolution. *Land Use Policy*, *87*, 104065.

Simonofski, A., et al. (2017) Citizen participation in smart cities: Evaluation framework proposal. In *2017 IEEE 19th conference on business informatics (CBI)* (pp. 227–236). Thessaloniki, Greece.

Tomor, Z. et al. (2019). Smart governance for sustainable cities: Findings from a systematic literature review. *Journal of Urban Policy*, *26*(4), 3–27.

Ullah, F. et al. (2021). Risk management in sustainable smart cities governance: A TOE framework. *Technological Forecasting and Social Change*, *167*, 120743.

Valkenburg, A. C., den Ouden, P. H., & Schreurs, M. A. (2016). Designing a smart society: From smart cities to smart societies. In B. Salmelin (Ed.), *Open innovation 2.0 yearbook 2016* (pp. 87–92). European Commission.

Vanolo, A. (2016). Is there anybody out there? The place and role of citizens in tomorrow's smart cities. *Futures*, *82*, 26–36.
Yigitcanlar, T. (2015). Smart cities: An effective urban development and management model? *Australian Planner*, *52*(1), 27–34.
Yun, Y., & Lee, M. (2019). Smart city 4.0 from the perspective of open innovation. *Journal of Open Innovation: Technology, Market, and Complexity*, *5*(4), 92.
Zhao, F. et al. (2021). Smart city research: A holistic and state-of-the-art literature review. *Cities*, *119*, 103406.

INDEX

Pages in *italics* refer to figures and pages in **bold** refer to tables.

Kolibri 360 GmbH
Ferdinand-Jühlke-Straße 7
99095 Erfurt, Deutschland
produktsicherheit@kolibri360.de